INTERNATIONAL MINERALS CARTELS AND EMBARGOES

A Charles River Associates Research Report

Michael W. Klass
James C. Burrows
Steven D. Beggs

with

Bernard Reddy
Scott Cardell
Thomas Cooley
Gilbert DeBartolo
Frederick Dunbar
Robin Landis
Douglas Woods

INTERNATIONAL MINERALS CARTELS AND EMBARGOES

Policy Implications for the United States

PRAEGER

PRAEGER SPECIAL STUDIES • PRAEGER SCIENTIFIC

Library of Congress Cataloging in Publication Data

Klass, Michael W
International minerals cartels and embargoes.

"A Charles River Associates research report."
Includes bibliographical references and index.
1. Mineral industries. 2. Trusts, Industrial.
3. Strategic materials--United States. I. Burrows, James C., joint author. II. Beggs, Stephen, joint author. III. Charles River Associates. IV. Title.
HD9506.A2K58 380.1'42'0973 80-11123

ISBN 0-03-044366-0

Published in 1980 by Praeger Publishers
CBS Educational and Professional Publishing
A Division of CBS, Inc.
521 Fifth Avenue, New York, New York 10017 U.S.A.

0123456789 145 987654321

Printed in the United States of America

PREFACE

The continued existence of the petroleum cartel (OPEC), recent disruptions of petroleum supplies from Iran, and civil disturbances in Zaire, the source of much of the world's cobalt, highlight the industrial world's dependence on a relatively few suppliers of important materials. And, fears of cartel price gouging and politically motivated embargoes continue to provide a basis for much of U.S. energy policy. This book addresses the policy issues posed by such risks; it assesses the likelihood of cartels or embargoes involving a number of important materials, and then applies a consistent economic framework to evaluate a range of policies the United States might adopt in anticipation of such contingencies.

We believe that our effort supplements the contributions made by other studies, such as the Council on International Economic Policy's "Red Book," Critical Imported Materials, and, more recently, John Tilton's Brookings Institution study, The Future of Non-Fuel Minerals. Our work applies a consistent analytical framework to evaluate the economic merits of a wide range of economic policies, including stockpiling, development of alternative technologies, tariffs, and subsidies. It is industry specific: the framework is not simply a black box, but is implemented wherever possible with industry-specific information. While the book is rather sweeping in scope, covering not only petroleum, but six other major nonfuel minerals as well, it focuses primarily on the economic merits of the policies considered. We did not attempt to treat the full range of foreign and domestic policy considerations raised by the problem of potentially insecure imported materials.

This book is a product of research on mineral industries and markets begun at Charles River Associates (CRA) in 1966. The most direct source for the research effort leading to this book came from the Experimental Technology Incentives Program (ETIP) of the National Bureau of Standards, U.S. Department of Commerce. Under the direction of Dr. Jordan Lewis, ETIP sought an understanding of when the U.S. response to threatened supply interruptions or cartels should be a technological one, and when it should take other forms. ETIP contracted with Charles River Associates to conduct the research. The results of Contract 4-35960 were a 10-volume Charles River Associates report, Policy Implications of Producer Country Supply Restrictions, completed in 1976 and a conference sponsored by ETIP in January 1977. Volumes included a summary and overview, a discussion of the analytical framework, proceedings of the ETIP Conference on Commodities Supply Policies, and separate volumes on seven specific commodities: aluminum-bauxite, chrome, copper, cobalt, manganese, petroleum, and platinum. This volume is in large part a distillation, refinement, and summary of the findings and analysis of that study.

We are particularly indebted to Dr. Lewis and to Gregory C. Tassey, Chief of Economic Assistance Policy, for their support of the research effort and for their consistently helpful suggestions. The book draws extensively on research undertaken for the Department of Interior, under contract number 14-01-0001-2182, in the development of optimal stockpiling models (Charles River Associates, The Report of the U.S. Department of the Interior on the Critical Materials: Aluminum, Chromium, Platinum, and Palladium: A Review and Revision [Cambridge, Mass.: CRA, July 1977]). The present book also draws on research undertaken for the Bureau of International Labor Affairs of the Department of Labor under contract number ILAB 75-3. This resulted in a report entitled, Economic Issues Underlying Supply Access Agreements: A General Analysis and Prospects in Ten Mineral Markets (Cambridge, Mass.: CRA, July 1975). Dr. Harry E. Grubert, then Director of Economic Research for the Bureau of International Labor Affairs, suggested the study and provided helpful suggestions. Finally, the analysis draws on related work on mineral policy problems undertaken for the National Commission on Supplies and Shortages, which resulted in a report entitled, Private and Public Stockpiling for Future Shortages (Cambridge, Mass.: CRA, August 1976). The latter work benefited from the suggestions of Dr. George Eads, Executive Director of the National Commission on Supplies and Shortages, and of Dr. Alan Cohen of his staff.

A number of writers contributed to the research reported in this book. Michael W. Klass, then of Charles River Associates, oversaw the research work for ETIP and the Department of Labor, including development of the analytical framework, and was the principal author of this book. James Burrows had overall responsibility for the project and contributed to the research in a number of areas, particularly bauxite, cobalt, and energy. Steven Beggs of Charles River Associates developed portions of the analytical framework and performed basic analysis of chromium and manganese. Bernard Reddy of CRA coordinated much of the updating of the results. Sections on particular commodities are based on research by the following: Thomas Cooley—platinum; Gilbert DeBartolo—copper; Frederick Dunbar—cobalt; Robin Landis—petroleum; and Douglas Woods—aluminum/bauxite. Others on the CRA staff made notable contributions. N. Scott Cardell programmed the most general version of the CRA policy model, described in the appendix of this study. Timothy Greene performed much of the analysis of energy alternatives and of platinum. David Blau contributed greatly to the research on chromium, as did Dianne Hederich for manganese. Marilyn Edling of CRA and Paul Eckbo of MIT contributed to the work on energy. Several academic consultants contributed substantially to the study. Professor John Elliott of MIT contributed to the analysis of chromium and particularly manganese. Professor Steven Krasner performed work related to the past history and success of cartels as well as policies toward them. Professors Morris Adelman and Franklin Fisher of MIT and Zvi Griliches of Harvard reviewed various portions of the underlying research and made helpful suggestions. Dr. John Morgan of the U.S. Bureau of Mines and many of the Bureau's experts offered many useful comments on the final CRA report. Finally, special thanks go again to Dr. Lewis and to Greg Tassey, of ETIP, as well as to the members of the ETIP review panel, including Professors Harold Barnett and William Vogely, James Owens, James Miller, Gordon Smith, Nick Scodari, Emmanuel Horowitz, Donald Jordan, Bruce L. Robinson, and Samuel Rosenblatt.

The above listing of helpers and contributors is not an attempt to implicate others. The views expressed in this book are only those of the principal authors, and the responsibility for errors, of course, rests with us. The views expressed in this book do not necessarily reflect the position of the Federal Trade Commission or any individual commissioner, or of the Experimental Technology Incentives Program or any other government agency, or Charles River Associates.

This book is a summary of the findings and analysis of the 1976 study, and has not been updated to reflect 1980 conditions. The major changes since 1976 have been in the energy market, with the determination in the security of supply of Middle Eastern oil proving much graver threats of supply interruptions than was the case in the mid-1970s. Since our study was completed the Department of Energy has been formed, numerous studies and monographs on energy policies have been published, the Carter Administration has developed a new generation of energy policies, including the creation of the Energy Mobilization Board of the Energy Security Commission. These latter bodies would expedite the implementation of the type of technology of policies assumed in this book (such as coal conversion and nuclear power). Conservation policies, which were beyond the scope of our original study, have received great attention in recent years, and numerous analysts have demonstrated the cost-effectiveness of specific conservation policies. The energy portion of this book should therefore not be interpreted as providing an integrated assessment of current energy problems and policies, but rather as illustrating methodological approaches to certain generic policy questions.

CONTENTS

Chapter Page

LIST OF TABLES AND FIGURES

Table		Page

Figure		Page

PART I

OVERVIEW, POLICY CONCLUSIONS, AND RISK ASSESSMENTS

1
OVERVIEW

INTRODUCTION

Between 1972 and 1974 the industrialized world suffered a number of traumatic shocks that raised the issue among some of the possibility of a new world economic order. Beginning in late 1972, both agricultural and mineral prices began to rise rapidly. According to The Economist magazine's indexes, by the time the commodity price boom had peaked, world open-market agricultural food prices had risen by 267 percent and mineral prices by 240 percent from their early 1970s lows. The October 1973 Middle East war brought the Arab oil embargo, which then triggered an eightfold increase in OPEC taxes and a quadrupling in f.o.b. Middle Eastern oil prices. The International Bauxite Association was formed, and bauxite taxes levied by Jamaica and other countries were increased from about $1.80 per ton to about $15.00, causing delivered U.S. costs of bauxite to approximately double. Producer associations in copper, tungsten, mercury, and iron ore were formed or became more active. Finally, the less developed countries began pressing in international organizations for fundamental changes in the way minerals are marketed.

The OPEC embargo and price increases, the commodity price boom and inflation of 1972-74, and the demands of less developed and other supplying countries for a new international economic order led to increasing concern over the proper role of the federal government in contingency planning for possible crises in supplies of basic commodities. Proposals have ranged from economic stockpiling to federal funding and development of advanced technologies that would lessen import dependence and ease the impact of possible supply crises, to general central planning of materials supply and demand.

SCOPE OF THE ANALYSIS

This book examines a major issue in materials policy: proper federal policies toward supply restrictions by countries that produce important raw materials imported by the United States. The scope of the analysis is both unusually broad and necessarily circumscribed. The breadth comes from the need for a realistic and solidly based evaluation of a wide range of policy options, both technological and nontechnological. Such an evaluation must be both conceptually general and empirically market specific, and must use a wide range of research techniques. For concreteness, specific markets were studied: aluminum-bauxite, chromite, cobalt, copper, manganese, energy, and platinum and palladium. These markets display a wide range of import dependence, producer-country characteristics, ease of substitution, and technological issues.

We did not address the full range of issues in materials policy, which includes resource exhaustion, price stability and investment adequacy, temporary shortages, commodity agreements and buffer stocks, access to supply agreements, and producer-country control of extractive industries; full investigation of these issues was beyond the scope of this project. Neither did we concentrate on potential restrictions due to wars or natural disasters, nor did we treat the impact of supply restrictions on national defense or security. Rather, we focused on deliberate producer-country supply restrictions, whether concerted or unilateral, treating issues such as resource exhaustion or adequacy only as they related to this major concern. Much of the analysis is, of course, relevant to restrictions due to causes other than deliberate producer-country actions.

We addressed a range of policies, rather than giving exclusive or primary emphasis to any one policy instrument. The principal options we considered are stockpiling, tariffs (and quotas), and subsidies, both general and specific technological options. We have concentrated on these policies for several reasons: they have been widely proposed and discussed, they can directly affect import dependence and the impact of restrictions, and they are amenable to economic analysis. Our focus was on more directly economic or market-oriented policies.[1]

The scope of the book is limited to evaluating risks and policies from the U.S. point of view; other countries may face different risks or policy options, but they are not a major concern of this study. The time horizon for the analysis was 1990, so we considered only contingencies and technological options judged likely to be relevant by that date. In addition, much of the research was completed in 1976, and data on many topics are current as of that date. The analysis has not been updated to 1980. Much of the analysis, particularly of energy, analyzes "1980 scenarios" and refers to 1980 as if it were several years hence. Where this is true we urge readers to read "relatively near term" for 1980 and "well over ten years hence" for 1990. The specifics of the timing and results are, we believe, less significant than the methods of analysis and the general conclusions.

While a major focus of the analysis was the usefulness of technological change as an instrument of contingency planning, original technological forecasting was beyond our scope. Rather, we concentrated on the usefulness of available and known technologies and on the general potential of technological change as compared with a variety of other policy instruments. This effort, combining market specifics and realistic assessments of risks and impacts with analysis of a range of policy alternatives, is intended as a model of the type of analysis needed for national policy making and as a guide to the kinds of additional research and analysis needed for specific policy decisions.

APPROACH TO THE ANALYSIS

A major premise of our study was that rational policy analysis must proceed from a realistic and market-specific analysis of risks to a similarly concrete evaluation of potential damage, and then to consideration of a range of policy instruments including not only government encouragement of technological change, but also stockpiling, tariffs and quotas, subsidies, and other measures.

Our approach combined an assessment of real dangers to the U.S. economy in various materials markets with an analysis of the costs of different policies designed to reduce the impacts of foreign supply restrictions. The assessment of the dangers involved an analysis of the economic costs of various possible contingencies and an assessment of the probabilities of these contingencies. The analysis of the potential policies also involved the measurement of the transfer payments (income redistributions) and the identification of the noneconomic costs, such as environmental consequences, of each policy.

The benefits of each policy were defined to be the present discounted value of the probable reduction in cost impacts arising from possible policy responses to future supply restrictions. The benefits of most policies are uncertain, as the benefits occur only if a supply restriction occurs. Therefore, the benefits are weighted by assessments of the probabilities of each contingency. The present discounted value of the benefits was computed to take into account the fact that benefits that occur in the near term are worth more than benefits that occur in the distant future. For each policy, the present discounted value of the expected benefits was then compared to the present discounted value of that policy's costs, to determine whether the policy was justified on economic grounds. By quantifying the costs and benefits of policies in this manner, we attempt to give policy makers some of the material with which to make informed decisions in cases where noneconomic policy goals are involved.

There are many policies which might be recommended on narrowly economic grounds, but which would create such undesirable side effects that they would generally be ruled out. For example, if it were known in advance that another oil embargo were going to occur in 1980, in the absence of other controls, an oil-import quota of economically optimal size would yield sizable economic benefits, on the order of $13.7 billion. However, the quota could, in the absence of carefully designed offsetting policies, also lead to windfall profits to producers of slightly over $100 billion.

FINDINGS

The study analyzed in detail the seven import-dependent mineral markets generally considered to be of the greatest national concern: aluminum/bauxite, chromium, manganese, cobalt, platinum/palladium, copper, and petroleum. Table 1.1 presents salient data on these markets.

It is difficult to generalize about the results reached for different markets; in fact, perhaps one of the most important conclusions of our study is that each market must be analyzed on its own terms. It does not follow that high import dependence, concentration of supply, or the existence of ongoing cartel efforts necessarily imply high risks for the U.S. economy.

In general, it appears that on the one hand, the United States faces substantial risks of costly producer-country supply restrictions in a number of important nonfuel mineral markets, including chromite, cobalt, platinum, palladium, manganese, and tin (not studied). On the other hand, the likelihood of traumatic supply restrictions in most other mineral markets seems small. This is particularly true for many of the most important markets: iron and steel, copper, nickel, molybdenum, lead, zinc, magnesium, and tungsten.

One must further distinguish between risks of short-run disruptions (such as embargoes) and of long-run price increases from cartelization. Short-run supply cutoffs are quite possible for chromite, cobalt, platinum and palladium, and possibly for manganese. Such cutoffs would be extremely costly—for example, in the cases of chromite, cobalt, and manganese, price increases of as much as eight- or ten-fold could conceivably occur; however, private stocks in these markets tend to be substantial and the presence of additional public stockpiles would enable the United States to bridge the shortfall in supplies for some time.

Damaging cutoffs seem unlikely in the cases of bauxite and copper. In the case of bauxite, an export embargo by a producer as large as Jamaica would be costly to the United States, but in the long run we would adjust toward other supply sources, including U.S. nonbauxitic aluminum-bearing ores. As Jamaica is well aware that its bauxite reserves are not unique, it is difficult to imagine Jamaica actually implementing a protracted cutoff for economic reasons alone. Nevertheless, the high costs of such a cutoff lead us to suggest maintenance of a contingency stockpile of bauxite. In the case of copper, the United States has sufficient domestic resources to insure that no one supplier could cause irreparable damage; and suppliers must be concerned about long-term losses in markets as a result of precipitous supply actions. Increasing import-dependence could expose the United States to the risks posed by supply interruption in the future.

The likelihood of long-run price increases from current levels as the result of cartel actions appears to be most significant in the cases of chromite and perhaps bauxite. In the case of bauxite, the International Bauxite Association (IBA) has already increased its taxes by a factor of 8, resulting approximately in a doubling of the U.S. cost of bauxite. The existence of domestic nonbauxitic sources of aluminum provides a potential substitute for imported bauxite. Available estimates of the costs of producing from these resources suggest that IBA taxes are already near the highest level that can be attained in the long run without major losses in the market for IBA bauxite; at most, IBA taxes would double if the higher cost estimates turn out to be realistic. Such a cost increase would be economically damaging to the United States, but it would not be traumatic. Furthermore, it appears, on close examination, that the IBA may not be able to enforce the 1976-77 tax structure; export taxes of the high-tax countries, such as Jamaica, may decline, at least in real terms.

In the case of chromite, there are only very limited supply possibilities, at historical price levels, outside of South Africa, Rhodesia, Turkey, and the Soviet Union. Chromite prices as of 1976-77 may well be below potential monopoly prices, and it appears that successful collusion among South Africa, Rhodesia, the Soviet Union, and Turkey could result in at least a doubling and perhaps a quadrupling of current prices.

TABLE 1.1

U.S. Import Dependence in Seven Critical Materials Markets

Market	Imports as Percent of Primary Market[a]	Imports (millions of 1974 dollars)	U.S. Import Shares		Non-Communist-World Supply Shares[b]		Uses
Aluminum/bauxite	91	622.4	Jamaica	40	Australia	29	motor vehicles,
			Australia	25	Jamaica	22	airplanes,
			Surinam	16	Surinam	10	construction,
			Guinea	5	Guinea	9	containers,
			Guyana	5	Guyana	4	electrical uses
Chromium	100	83.8	South Africa	27	South Africa	33	stainless steel:
			Rhodesia	13	Soviet Union[d]	19	steam turbines,
			Yugoslavia[c]	13	Turkey	11	motor vehicles,
			Soviet Union	11	Rhodesia	10	chemical and
			Philippines	7	Philippines	7	food processing
			Turkey	6	India	6	equipment
Manganese	100	133.5	France[e]	32	South Africa	31	all steels
			South Africa	20	Gabon	17	
			Brazil	12	Brazil	15	
			Gabon	10	Australia	12	
			Australia	6	India	12	

Cobalt	100	54.2	Zaire	45	Zaire	67	jet engines, magnets,
			Belgium/ Luxemburg[f]	37	Canada	8	metal-working
			Norway[g]	8	Morocco	6	machinery
Platinum/palladium	100	502.7	South Africa	31	Soviet Union	47	catalysts:
			Soviet Union	31	South Africa	46	motor vehicles,
			United Kingdom[h]	23	Canada (shares in world production)	6	petroleum refining, chemical production, electrical parts
Petroleum	37	28,000.0	Canada	23	United States	18	heating,
			Nigeria	20	Saudi Arabia	18	electricity
			Iran	13	Iran	13	generation,
			Saudi Arabia	13	Venezuela	6	chemicals
			Venezuela	9	Kuwait	5	
			Indonesia	8	Nigeria	5	
Copper	27	1,034.1	Chile	24	United States	23	electrical
			Canada	21	Chile	14	equipment,
			Peru	19	Canada	13	construction,
			Japan	13	Zambia	11	machinery
			South Africa	8	Zaire	9	
					Australia	4	
					Philippines	4	
					Peru	3	

(continued)

Table 1.1, continued

[a]Imports as a percentage of the sum of domestic primary production plus imports.
[b]Shares in sum of non-Communist production plus exports from Communist to non-Communist countries.
[c]Ferrochromium produced from ores mined in other countries.
[d]Exports to non-Communist world.
[e]Ferromanganese produced from ores mined in Gabon.
[f]Cobalt metal refined from African ores.
[g]Cobalt metal refined from Canadian materials.
[h]Platinum-group metals refined from Canadian and South African materials.

Notes: Import share = value of imports from country ÷ total value of imports.

Aluminum/bauxite

Import percentage $= \frac{2.09 * \text{alumina imports} + \text{bauxite imports}}{\text{imports} + \text{consumption of domestic ore}}$

$= \frac{2.09 * 3.627 + 14.308 * 1.12 + .304}{2.09 * 3.627 + 14.308 * 1.12 + .304 + 1.989 * 1.12}$

$= .91$

Import value = alumina import value + \$23.21 per long ton* bauxite imports + calcined bauxite import value
= \$270.617 + \$23.21* 14.308 + 19.732 = \$622.438

Chromium

Import value = value of chromite imports + value of ferrochromium imports (excludes silicochromium, chromium metal, and scrap)
= \$28.532 + \$22.127 + \$33.134 = \$83.8

Manganese

Import value = value of ore imports + value of ferromanganese imports (excludes silicomanganese, manganese metal, and scrap)
= \$45.091 + \$88.426 = \$133.5

Cobalt

Import value = value of metal and oxide imports = \$54.2

Platinum/palladium

Import value = value of unwrought and semimanufactured platinum-group metal imports
= \$502.7

Petroleum

$$\text{Import percentage} = \frac{\text{imports of crude oil and refined products}}{\text{consumption of refined products}}$$

= 2,231/6,078 = .37

Import value = average price paid by refiners for imported crude
* imports of crude oil and refined products
= \$12,44*2,231 = \$27,754

Copper

Copper imports = copper contained in imports of ore, concentrates, blister, and metal (excluding scrap)
= 577.444

$$\text{Import percentage} = \frac{\text{copper imports}}{\text{copper imports + domestic mine production}}$$

$$= \frac{577.444}{577.444 + 1{,}597.002} = .27$$

Import value = value of total copper imports
- value of copper-scrap imports
= \$1,084.793 = 50.717 = \$1,034.1

Sources: Except for petroleum, data are from U.S. Bureau of Mines, 1974 Minerals Yearbook, preprints (Washington, D.C.: Bureau of Mines, 1974). Petroleum data are from Mineral Industry Survey: Monthly Petroleum Statement, December 1974; Monthly Energy Review (Washington, D.C.: Federal Energy Administration), various issues; Charles River Associates, Policy Implications of Producer Country Supply Restrictions: The World Energy Market (Cambridge, Mass.: CRA, 1976).

Manganese supplies are somewhat more dispersed than chromite supplies, but the non-Communist world is becoming increasingly dependent on three suppliers: South Africa, Gabon, and Australia. South Africa has the largest manganese resources outside of the Soviet Union, but production costs are rising and may ultimately be two to three times the levels of 1976-77. A disruption in manganese supplies would be very costly, and contingency policies should be designed to deter such disruptions and to moderate their impact should they occur. The present importance of Australia and Brazil as suppliers makes explicit cartelization unlikely, but manganese prices may well rise over time.

In the cases of platinum and palladium, it appears that producing countries are already charging a price that is near the monopoly optimum. Further price rises would of course be costly to the United States and other consumers, but they would also be costly to the suppliers, as such increases would result in a long-run decline in net revenues. The high concentration of supply in these markets makes supplies uncertain, but also makes further, deliberately traumatic price increases unlikely.

Cobalt is already monopolized by Zaire, and cartelization is therefore not an important contingency. Supplies in Zaire are insecure, as evidenced by the effects of guerrilla activities there in 1976, 1977, and 1978. The costs of such disruptions could be severe, and stockpile policies to moderate the impacts of disruptions may be highly desirable.

Finally, cartel-induced, protracted price increases seem unlikely in the copper market. The countries which seem likely to engage actively in cartelization attempts simply do not have enough leverage on the market to enforce substantial and protracted price increases. However, the United States will become increasingly import-dependent in copper and future supply interruptions may become a serious problem.

In the energy market, OPEC's collapse does not appear imminent and oil prices can be expected to continue to increase at the rate of inflation or faster. U.S. import dependence will continue to be substantial, and there is significant risk of future supply cutbacks by OPEC to serve political ends. The disruptions considered in this book are more severe than those of 1973-74 and 1979, but seem mild in comparison to the disruptions regarded as being at least possible by analysts today (such as the total loss of Persian Gulf supplies for an indefinite period). Recent events in Iran highlight our dependence on politically volatile suppliers.

Even for the "moderate" disruptions analyzed, an economic-contingency stockpile equaling at least two of three months' U.S. consumption is justified. Substantially greater stockpiles would be justified for such severe contingencies. Tariffs, quotas, and general subsidies are probably not justified, in part because of the large income redistributions such measures would entail. Acceleration of coal conversion for fossil-fueled electric-power generation, assuming that environmental problems can be overcome, seems to be clearly cost effective, but the likely oil-import displacement is not substantial relative to current oil imports. Acceleration of nuclear power plant construction also seems justified on narrow economic-cost grounds, although the oil displacement would be minimal. Finally, the costs of oil shale and low-Btu gas appear to be low enough relative to current oil prices to warrant some moderage government incentives. Other synthetic-fuel technologies seem to be too costly to justify government support other than further research and development and some pilot plant testing. Analysis of conservation policies was beyond the scope of our study, though conservation policies may be the most cost-effective means of dealing with the costs and risks imposed by OPEC.

Technology Policies

In general, subsidizing new technology was not found to be cost effective; in most cases stockpiling tended to be a much less expensive policy instrument for reducing the impact of supply interruptions. Subsidizing new technology also does not seem to be cost effective for combating long-run price increases from cartelization activities. For example, in the case of aluminum there is no economic gain to the United States from producing alumina from domestic nonbauxitic ores at a bauxite-equivalent cost of $40 as long as bauxite can be imported at $25 a ton, even if the latter figure includes $10 in monopoly profits to the producing countries.

On the other hand, there are numerous instances in which selective technological policies can have substantial contingency benefits. For example, research and development on reducing the costs of producing alumina from domestic nonbauxitic ore might reduce the IBA's price and otherwise induce the cartel to moderate its policies. There is a potential for a legitimate public role in such efforts, particularly if the new technologies are not likely to be profitable under normal market conditions.

Another instance in which government involvement in new technologies may be justified may occur when technologies are commercially viable or nearly commercially viable at cartel prices, but would not be commercially viable at lower prices. Examples of this may be shale oil and low-Btu gas from coal, whose costs appear to be low enough to be very near commercial viability at current oil prices; the cost estimates, however, are highly uncertain. To the extent that such technologies are being delayed because of concerns that oil prices may decline in real terms, there may be justification for offering price supports at levels below current market prices to reduce the risks of investment. The desirability of such a policy clearly depends on the risk assessments. If in fact there is a substantial probability that oil prices will decline, such a price-support policy could prove to be very costly and inefficient. On the other hand, if future embargoes are considered highly likely, if the probability of OPEC's collapsing from normal market forces is regarded as small, and if there is some probability that encouraging new energy technologies may increase the probability of an OPEC collapse, then selective government incentives for certain new technologies may be cost effective.

Other examples of technological policies we identified as being potentially desirable include:

1. government programs to insure wide access to information on methods of manganese and chromium conservation;
2. expedited adjustment of manganese and chromium specification ranges for steel alloys during emergencies;
3. encouragement of ocean mining, which could have favorable contingency benefits for copper, cobalt, manganese, and nickel (not studied);
4. dissemination of information on nonplatinum catalysts; and
5. research on reducing the use of stainless steel through cladding and other techniques.

Contingency Stockpiling

Stockpiling is a particularly efficient form of protection against supply interruptions, and stockpiles are a very useful policy instrument to deter cartel attempts and to moderate the impacts of cartels once they have been formed. Three broad types of stockpiles can be distinguished: military stockpiles, buffer stockpiles, and economic stockpiles. The existing national strategic stockpile is maintained for military contingencies and emergencies, and the materials held to satisfy the stockpile goals established under the Strategic and Materials Stockpiling Act of 1946 (which was expected to be amended by 1979) are not available for nonmilitary emergencies. Materials held in surplus of stockpile objectives (current goals) can be disposed of, subject to congressional authorization.

The United States traditionally has not maintained stockpiles of minerals and other strategic materials for price-stabilization purposes, but at times surplus strategic stocks have been disposed of in a manner which seems at least partly motivated by a drive to stabilize prices. More recently, the United States has agreed to contribute 5,000 tons of tin to the International Tin Agreement buffer stock, and has been engaged in negotiations to establish international commodity agreements (which may include plans for buffer stocks) in copper, tungsten, and several other mineral commodities.

Contingency stockpiles that could be used in peacetime emergencies (which might be caused, for example, by a disruption in supplies from central and southern Africa) are conceptually similar to buffer stocks, in that both contingency stocks and buffer stocks are designed to prevent major price swings. Although a practical distinction between contingency and buffer stocks may be difficult to implement, contingency stocks, as described in this book, would be designed to prevent massive price changes caused by major disruptions (such as the recent tenfold increase in open-market cobalt prices resulting from the invasion of Shaba Province in Zaire). More normal price swings resulting from the business cycle and normal changes in demand and supply would presumably not be addressed by economic-contingency stocks.

In October 1976, the strategic-stockpile goals for most materials in the U.S. strategic stockpile were substantially increased, reflecting a decision to hold stocks for a three-year (instead of a one-year) military emergency. As of March 31, 1978, the value, at existing market prices, of the 93 materials in the U.S. strategic stockpile was $9 billion (of which $3.9 billion worth was held in excess of stockpile goals); the total market value of the stockpile goal amounts was substantially greater than $9 billion. Table 1.2 summarizes new stockpile goals and inventories, as of October 1, 1976, for the nonfuel materials analyzed in this book. The information in Table 1.2 is reduced to broad commodity categories in Table 1.3 for ease of comparison with the economically efficient stockpile levels presented in this book. Table 1.3 also presents data on 1976 U.S. consumption of each commodity, for comparison with the stockpile levels.

We found that contingency stocks for peacetime disruptions—in addition to the existing stockpile of strategic and critical materials—are justified for all the markets studied, with the possible exception of copper. The illustrative ranges for economically efficient stockpile levels presented in the chapters on each commodity are as follows (in months of U.S. consumption):

Commodity	Months
Bauxite	6 to 12
Chromium	18 to 24
Cobalt	8 to 20
Copper	0 to 2
Manganese	12 to 18
Platinum and palladium	2 to 12

TABLE 1.2

U.S. Strategic and Critical Materials Stockpile Goals and Inventories as of October 1, 1976

Material	Unit	New Stockpile Goal	Old Stockpile Goal	Total Government Inventory	
				Quantity	Millions of Dollars
Aluminum					
Alumina	st[a]	11,532,000	0	0	0
Aluminum	st	0	0	5,704	5.476
Aluminum oxide, fused crude	st	147,615	0	249,009	44.948
Aluminum oxide, abrasive grain	st	75,000	17,200	50,905	15.781
Bauxite, Jamaican	ldt[b]	423,000	4,638,000	8,858,881	213.942
Bauxite, Surinamese	ldt	0	0	5,300,000	153.170
Bauxite, refractory	lct[c]	2,083,000	0	173,000	20.414
Chromium					
Chromite, chemical	st	734,000	8,400,000	250,000	12.722
Chromite, metallurgical	st	2,550,000	444,710	1,952,802	256.305
Chromium, ferro, high carbon	st	236,000	11,476	402,001	318.987
Chromium, ferro, low carbon	st	124,000	0	298,054	379.054
Chromium, ferro, silicon	st	69,000	0	55,608	40.010
Chromium, metal	st	10,000	0	3,763	18.363
Chromite, refractory	st	642,000	54,000	399,960	25.309

Cobalt	pounds	85,415,000	11,945,000	40,693,169	164.415
Copper	st	1,299,000	0	0	0
Manganese, battery, natural	st	12,736	10,700	209,684	28.820
Manganese, battery, synthetic	st	19,105	0	3,008	1.445
Manganese ore, chemical	st	247,136	12,800	145,586	9.500
Manganese ore, metallurgical	st	2,052,000	750,000	3,145,946	226.613
Manganese, ferro, high carbon	st	439,000	200,000	600,000	227.676
Manganese, ferro, low carbon	st	0	0	0	0
Manganese, ferro, medium carbon	st	99,000	10,500	28,921	19.446
Manganese, ferrosilicon	st	81,000	15,900	23,574	10.962
Manganese, metal, electrolytic	st	15,000	4,750	14,166	15.300
Palladium	to[d]	2,450,000	328,500	1,254,994	72.613
Platinum	to	1,314,000	287,500	452,645	79.213

[a]Short ton.
[b]Long dry ton.
[c]Long calcined ton.
[d]Troy ounces.
Source: Metals Week, October 11, 1976.

TABLE 1.3

Summary of U.S. Strategic and Critical Materials: Goals and Inventories Compared to U.S. Consumption, for Basic Materials Markets (calendar year 1976)

Material	Units	Total Stockpile Inventory	Total Stockpile Goal	Stockpile Excess (Deficit)	1976 U.S. Consumption
Aluminum	millions of short dry tons of bauxite equivalent	16.7	26.3	(9.6)	22.462
Chromium	millions of short dry tons of contained chromium in ore equivalent	1.42	1.56	(0.14)	0.53
Cobalt	millions of pounds	40.693	85.415	(44.722)	16.482
Copper	millions of short tons	0	1.299	(1.299)	1.992
Manganese	millions of short tons of contained manganese in ore equivalent	1.99	1.49	0.50	1.36
Palladium	millions of troy ounces	1.255	2.450	(1.195)	0.657
Platinum	millions of troy ounces	0.453	1.314	(0.861)	0.851

Note: Stockpile inventory and goal as of October 1, 1976.

Sources: Metals Week, October 11, 1976; U.S. Bureau of Mines, Minerals Yearbook, Chapters on Aluminum, Chromium, Cobalt, Copper, Manganese and Platinum/Palladium Group Metals (Washington, D.C.: Bureau of Mines, 1976); American Metal Market, Metal Statistics (New York: Fairchild Publications, 1978).

Our findings should not be interpreted as suggesting that public contingency stockpiles within the ranges given here should in fact be established. First, there are enormous practical problems in managing contingency stocks. For example, without well-defined rules, the stockpile managers would almost certainly be subjected to intense pressures from special-interest groups to sell or purchase materials at times and in a manner which might not be in the national interest. Second, efficient stockpiles depend heavily on subjective judgment about the types and likelihoods of possible contingencies; the development of these assumptions is in the domain of the policy maker. Third, private industry typically carries substantial stocks in many critical markets, and in some cases private stocks may already be adequate. There is also reason to believe that public stocks substitute for private stocks, and that the quantities which private firms stockpile will be affected if the government initiates stockpiling or increases the level of its stocks.[2]

Tariffs, Quotas, and Subsidies

For the markets we studied, tariffs, quotas, and general subsidies generally did not seem to be desirable policies. In most cases the responsiveness of demand and supply to price is relatively small. To reduce import dependence substantially with such indirect policy instruments would entail large resource costs and inefficiencies in nonemergency years. Also, the implied income redistributions to domestic producers are enormous compared to the potential benefits of the policies.

Summary

For the nonfuel minerals, risks of producer-country supply interruptions seem to be concentrated in a few markets, most notably, chromite, cobalt, manganese, bauxite, platinum, and palladium. The costs of a major disruption in any one of these markets would be substantial relative to the size of the markets, but disruptions of the size we consider likely would not have a traumatic or crippling effect on the entire economy. The most effective policy to offset the contingency risks appears to be stockpiling; the benefits of most other policies are outweighed by their costs.

NOTES

1. For a discussion of other policies, see Charles River Associates, _Policy Implications of Producer Country Supply Restrictions: The World Energy Market_, vol. 2, "A Framework for Analysis" (Cambridge, Mass.: CRA, 1976).

2. See Charles River Associates, Public and Private Stockpiling for Future Contingencies, prepared for the National Commission on Supplies and Shortages (Cambridge, Mass.: CRA, August 1976).

2
THE FRAMEWORK OF ANALYSIS AND GENERAL POLICY CONCLUSIONS

One goal was to evaluate a range of technological and non-technological policy options, both in general and in specific market contexts. A major premise of our effort was that rational policy analysis must proceed from a realistic and market-specific analysis of risks to a similarly concrete evaluation of potential damage, and then to consideration of a range of policy instruments including not only government encouragement of technological change, but also stockpiling, tariffs and quotas, subsidies, and other measures. Figure 2.1 illustrates the sequence of analysis. Risks clearly depend on the market situation and political alignments; impacts depend on the alternatives available to the United States, both on supply and demand sides. Policy options are diverse and costly, so choice must be based on risks and potential impacts; the choice of policy can affect both the risk and the likely impact.

We first developed an analytical framework for evaluating the likelihood and impact of supply restrictions implemented by producer countries, the potential adequacy of private-sector actions in dealing with restrictions, and the merits of alternative federal policies. The framework was then applied to past and potential future crises in seven important markets: aluminum/bauxite, chromite, cobalt, copper, energy, manganese, and platinum/palladium.

To implement the framework, we relied on published information, our own background in materials research, contacts with industry and other experts, and extensive original research. Our methods included conceptual economic analysis, historical research, econometric modeling and simulation, and engineering-economic evaluations. Using such methods, we attempted to develop estimates of the likelihood of economically or politically motivated

FIGURE 2.1

Components of Policy Analysis

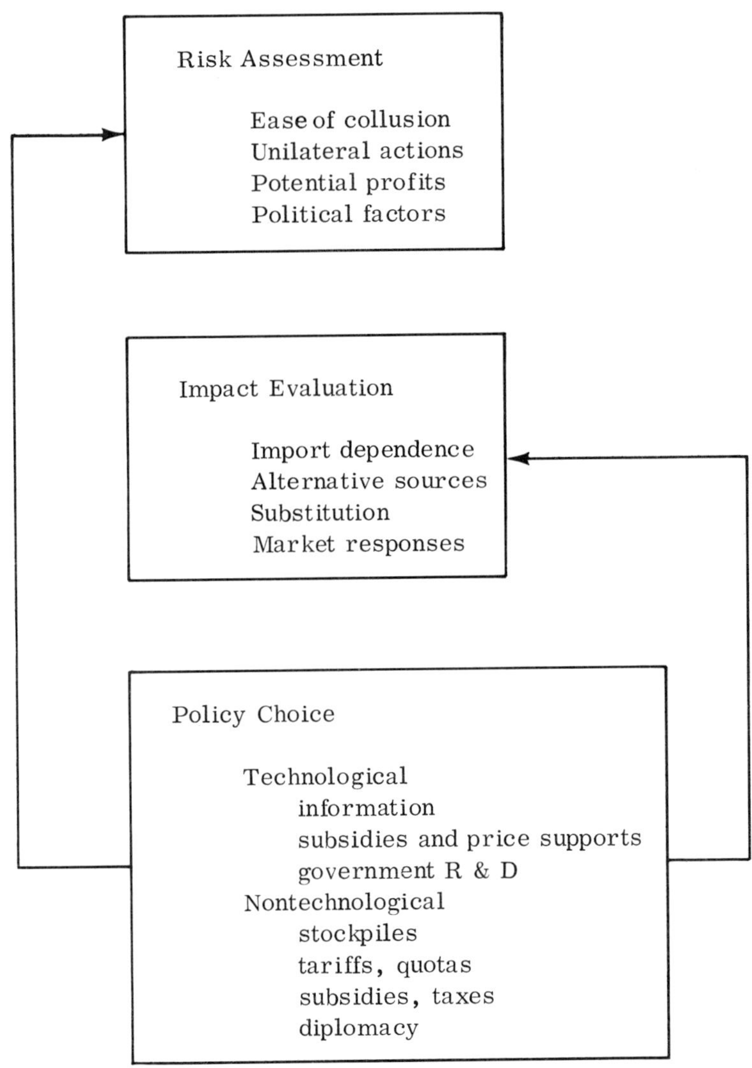

Source: Charles River Associates, Policy Implications of Producer Country Supply Restrictions: Overview and Summary (Cambridge, Mass.: CRA, 1976), p. 6.

disruptions in materials supplies or of escalations of prices that could substantially damage the United States.

THE RISK OF SUPPLY RESTRICTIONS

Doomsday scenarios of widespread and frequent materials embargoes and dramatic price escalations are unsupported by current or future market realities; in most cases the United States is not without alternatives. The task of the policy maker is not to prepare for autarky, but to provide appropriate insurance against possible supply restrictions. We did find that the United States faces significant risks in some markets: of higher prices for some imported materials, possibly over the long term, and of potential short-term disruptions in supplies, perhaps caused by exogenous political events. While the materials studied in this book do not cover all possible cases, they do represent a broad spectrum of the types of risks, potential costs, and policy alternatives facing the United States over the next few decades.

In bauxite, copper, and petroleum the United States faces ongoing cartels (the IBA, CIPEC, and OPEC, respectively) of varying degrees of cohesiveness and market control. Correct U.S. policy depends on the likely market impact and stability of these groups, their potential impact on the United States, the potential for extreme pricing actions or embargoes, and the availability and costs of turning to alternative sources of supply or substitute materials or inputs. In these cases, policy makers must also consider the impact of their actions on the behavior and stability of a cartel.

We found that the mere existence of a cartel in itself has no necessary implications for policy. In the copper market, not only does CIPEC have minimal market control, but the United States could turn to domestic resources and substitute materials at low to moderate cost. The power of the IBA is more substantial but is ultimately constrained by the availability of domestic resources. In both cases embargoes or extreme pricing actions appear unlikely. OPEC has demonstrated its market control and appears likely to maintain it for some time. In the absence of major disruptions in the Middle East, the price charged by OPEC is likely to remain roughly constant in real terms. Feasible and efficient policy options do not appear likely to decrease substantially U.S. dependence on OPEC for many years, and a major policy concern should be that of providing insurance against the possibility of future embargoes or supply cutbacks.

The other markets examined, chromite, cobalt, manganese, and platinum/palladium, do not have formal cartels or producers' associations. In these markets the fundamental question is the likelihood and possible impact of formal or informal collusive actions or unilateral restrictions implemented by single supplying countries, however motivated. We found the potential problems differ between cobalt and the platinum group on the one hand, and between chromite and manganese on the other. Market structure and available quantitative and qualitative evidence indicate that the markets for platinum and palladium and for cobalt are already largely monopolized. Though a cartel of African cobalt producers (Zaire, Zambia, and Morocco) is possible, its actions would in no sense precipitate a crisis in the cobalt market. However, unilateral actions by Zaire, the dominant cobalt producer, could have a major impact, particularly in the short run; supply interruptions caused by guerrilla actions or transportation difficulties would have equally severe effects.

The dominant producers of platinum and palladium, South Africa and the Soviet Union, are unlikely to collude, formally or overtly, to cooperate in restricting output. Nevertheless, the possibility of unilateral restrictions or embargoes does exist, as it does in any market dominated by few producers. In these two related markets the fundamental policy questions are the future reliability of present supplying countries and the possible damage due to short-run disruptions.

In the cases of chromite and manganese, the identity of the supplying countries suggests that explicit OPEC-type cartelization is fairly unlikely. However, tacit cooperation is quite possible and could lead to substantial price increases. Furthermore, supply disruptions over the next two decades, created by such countries as South Africa and Gabon (manganese), and by South Africa, Rhodesia, and the USSR (chromite), are quite possible, even perhaps likely, whether as a result of political, military, or economic factors. Such disruptions would have extremely severe impacts on the markets in question and would cause economic damage to the United States that could be measured in the billions of dollars. However, disruptions in these markets, unless unusually severe and protracted, would not have devastating economy-wide impacts.

The general conclusions to be drawn from our conceptual and case studies of the likelihood of restrictions include the following:

1. Concentration of production does not necessarily imply likely explicit OPEC-type collusion. Producing countries may be politically diverse and even antagonistic. Limited tacit collusion may be possible among international suppliers.

2. The existence of an explicit cartel does not imply a high degree of risk to the United States. Not all cartels have significant power, producer countries generally will not present a monolithic front, and U.S. alternatives vary among commodities.
3. The presence of countries with relatively undeveloped production potential often restricts producer power. Thus, U.S. policy might emphasize good economic and political relations with such countries and aim to encourage development of their potential.
4. Substantial restrictions may harm producer interests. Countries dominating their markets may already have taken advantage of their positions. Political, military, and trading ties with the United States and other consuming countries restrict freedom of action. Substitution, and entry and expansion of other producers, may damage the dominant producers' income and employment.
5. Market responses may constrain producer power. Private and public inventories constrain short-run power. Other restraints come from the entry or expansion of other producers, decreases in demand due to possibly irreversible substitutions, replacements, or conservation measures, increased recycling, and technical change.
6. Concerted restrictions are unstable. Prices above production costs give individual producers incentives to compete by expanding output, thereby undermining the restriction. An immediate policy implication is that consumer governments might attempt to increase competitive tendencies among alternative supply sources, rather than encouraging the producers' cooperation by dealing with them as a group.

Other findings indicate the following conditions for significant potential producer power:

1. Low-cost (or even all known) reserves and resources may be concentrated in a few countries. In our case studies this was particularly true for bauxite, chromite, cobalt, and platinum/palladium.
2. Domestic resources may be sparse or high in cost. This is particularly the case for chromite and for the platinum group. Resources are significant but high in cost in the cases of manganese and cobalt.
3. Development or expansion of alternative supplies may take substantial time or be constrained by bottlenecks. For example, expansions of South African chromite and manganese supplies in the 1970s have been constrained by transportation difficulties.

4. Substitution, conservation, or replacement may be costly or subject to technological uncertainty. We found this factor to be quite varied, but particularly present in manganese, in major uses of chromite, in platinum/palladium, and in petroleum.

ECONOMIC IMPACTS OF RESTRICTIONS

We applied an economic-surplus model to the evaluation of restrictions. The Appendix to this book presents an overview of the methods used to estimate costs and benefits. The model breaks down costs into three components: resource transfers to foreign producers due to higher prices; extra production costs due to replacement of some imports with previously higher-cost domestic alternatives; and losses to users due to decreased consumption and shifts to higher-cost production methods. We typically found that the resource-transfer component dominates the total cost except, of course, in cases such as total embargoes. The reason is clear: such costs are incurred on all imports persisting during the restriction. Other costs are incurred only to the extent that there are induced increases in domestic production or reductions in use.

A policy implication is that the costs of a restriction can be reduced in three basic ways: by reducing imports; by increasing the ease, and decreasing the cost, of substitution, replacement, or conservation; or by increasing the spead of adjustment to long-run production and consumption possibilities. Both technological and nontechnological policies can have impacts in all of these areas. Of course, the costs of policies themselves must also be weighed.

We also found that restrictions can have substantial domestic redistributive effects, conferring capital gains and increased incomes on domestic producers of the restricted commodity and its substitutes, and on workers in these industries; at the same time, such restrictions inflict losses on users of the restricted good. Significant employment decreases in using industries would occur only in two circumstances: if supplies, including inventories, were unavailable even at substantially elevated prices, or if price effects on final products were substantial enough to cause large decreases in total use. Both of these circumstances appear unlikely in the nonfuel minerals, at least for the cases we examined in detail. Because inventories generally are available, and imported-input cost represents a generally small proportion of final product price, except for oil, massive short-term decreases in consumption and employment are highly unlikely. The total employment effect of restrictions is rendered uncertain because of the likely offsetting increases in employment in the domestic industries producing substitutes and in related industries.

Our general finding is that the cost impact of restrictions diminishes as it becomes easier to shift away from imports, that is, as increasing domestic production or decreasing consumption becomes easier. And, of course, the impact of a restriction falls as the initial level of imports declines. Thus, estimation of restriction costs requires knowledge of the extent of import dependence and the costs of decreasing consumption and expanding domestic supplies, both in the short and long term.

Evaluating import dependence is not a trivial task, given the variety of forms in which many materials are imported, but it is not a major research task. Evaluating the costs of reducing consumption or increasing domestic production is substantially more difficult. In performing these vital tasks we relied on a variety of sources and techniques, including econometric estimation of demand and supply relationships; evaluation of supply potential drawn from studies of reserve and resource availabilities and costs; evaluation of the costs of alternative technologies for consumption and production; and discussions with industry and other experts.

Such a wide array of techniques was necessary to extract the greatest amount of meaningful information from scattered and often fragmentary data sources. For example, in the cases of manganese and metallurgical uses of chromite, statistical evidence suggests that demand is almost totally unresponsive to price changes. If this were in fact the case, that is, if users were willing to pay vastly inflated prices with no decrease in use, even a partial embargo could inflict enormous economic costs in the absence of stocks. Using a variety of nonstatistical methods combining engineering analysis with industry expertise, we found that flexibility in the use of the commodities was in fact feasible, though not without substantial costs. Statistical methods failed to disclose the underlying flexibility because past price and supply fluctuations had been too small, infrequent, or transient. Such results have an important implication for policy formation, namely, that accurate evaluation of the degree to which a material is in fact essential in given quantities must rest on careful examination of user technology.

Similarly, alternatives on the supply side cannot be appraised simply by examining past patterns, historical trends, or current practice. Rather, policymakers must assess a wide range of options, taking into account the technologies available or needed for their implementation. This leads us to another general conclusion of this study: the potential damage from a restriction will vary among commodities, depending on the underlying technologies and economics of supply and use. Therefore, no one policy measure or set of measures, whether technological or nontechnological, will be appropriate in all cases.

PROBLEMS OF PRIVATE-MARKET ANTICIPATION AND RESPONSE

We also evaluated the efficiency of the private market in anticipating and responding to restrictions, to provide a baseline for evaluating government policies and to indicate the proper role for government policy in this area. Our principal findings were that, while private-sector actions generally will be in the right direction for economic efficiency, major roles for government remain. Major potential sources of market failure in this regard include the following:

1. Firms may be unfamiliar with market or technological information that would be useful during a major disruption, due to a past history of price and supply stability.
2. Firms which make costly preparations for a restriction will be at a competitive disadvantage in the absence of a restriction, or if the restriction is brief.
3. Firms may engage in too little protective activity if their actions, in the aggregate, deter or moderate producer-country actions; because one firm's actions will confer benefits on other firms and on users, social benefits will exceed private benefits.
4. Firms expecting price controls, allocation schemes, or government-stockpile releases during disruptions will have reduced incentives to engage in protective activities.

All of these considerations indicate that private firms may engage in too little protective activity for economic efficiency; there is clearly a potential role for the federal government to design policies which are efficient from an overall national perspective. In addition, government policy makers may find that the private sector takes too little account of such factors as the effects of restrictions on other consuming countries, including Japan and Europe; the behavior of other groups of producing nations; the domestic distribution of income among regions, industries, or other divisions; and the international political and economic position of the United States.

Such considerations imply that government policy makers may find it desirable to take actions not dictated solely by economic efficiency. Our analysis concentrates on the economic impacts of restrictions and of alternative methods of federal contingency planning and response. While the political and social effects are surely important, detailed treatment of the extent to which policy measures should be oriented to them is beyond not only the scope of this project, but the purview of our expertise as economists.

POLICY TOOLS AND ALTERNATIVES

Our method for evaluating the economic impact of supply restrictions points to at least five ways in which policy measures can decrease the potential economic impact of restrictions. First, policy measures may decrease import dependence, either before or during a restriction, thus lowering the potential impact of restrictions. Import dependence may be decreased either by increasing domestic supplies, decreasing domestic demand, or both. Second, the cost of decreasing imports can be reduced by increasing the flexibility of domestic production, that is, by increasing the price elasticity of domestic supply. Third, the cost of import reduction can be reduced by increasing the flexibility of domestic use, that is, by increasing the price elasticity of domestic demand. Fourth, the speed with which alternatives may be implemented can be accelerated by increasing the rate of adjustment of domestic production and use toward their long-run possibilities. Finally, policies can reduce the likelihood, severity, or duration of restrictions; we say such policies yield "deterrence benefits."

While policy measures can yield benefits in the event of a restriction, they will also involve costs. These costs are of several types. There are costs of resources used as a result of the policy, such as investments in stockpiled resources, extra production facilities, or increased domestic production costs. The policies themselves will have administrative costs, and there are costs associated with foregoing low-cost imports in the absence of a restriction. There may also be incentive effects in private-market behavior; in particular, reduced private protective activities, such as stockpiling, may result from actual or expected government policy actions.

It is important to note that, in the absence of fundamental technological innovations, any policy decreasing import dependence moves the economy in the direction that a supply crisis would push it, and so imposes similar costs. Contingency-policy costs are incurred in the present and with some degree of certainty, while the benefits of such policies are uncertain (and may be zero if no restriction occurs, unless the policy itself deters the restriction).

A basic premise of this book is that comparison of policies' economic benefits and costs is a critical element in policymaking. Yet, measurement of these costs rests heavily on a subjective appraisal of risk and on judgments as to how heavily society should weigh possible effects. Further, both restrictions and contingency-planning measures have implications for domestic and international politics, for income distribution, and for other policy areas where criteria for optimality are even more uncertain.

In this context it is clear that economic cost-benefit analysis cannot itself provide definitive policy recommendations. However, the political and, even more directly, the distributional impacts of both restrictions and policies depend on the form and size of economic benefits and costs. Our findings and conclusions therefore provide essential information for the policy maker.

Benefits and Costs of Specific Policy Tools

Examining the sources of policy benefits and costs gives guidance about the market situations for which each policy tool may be most useful. This examination also lays a foundation for considering the question of substitutability and complementarity among policy tools, that is, the question of the appropriate mix of policies.

Information Provision

Government provision of market and technological information can facilitate private contingency planning and so lower restriction costs in any of the ways listed above. Costs include any administrative costs, costs of resources used in private planning, and possible decreased private-sector information gathering. Informational policies may have both short- and long-term usefulness. We found, for example, that knowledge of nonplatinum catalysts and methods for economizing on manganese use might be improved by a limited government effort.

Changes in Rules Affecting Demand, Supply, or Substitution Possibilities

Changes in policies that affect demand, supply, or substitution possibilities can yield benefits by lowering demand for imported materials, increasing domestic supplies, or enlarging the scope for substitution possibilities. Significant examples include building codes requiring the use of copper wiring; control of automotive emissions by platinum-bearing catalytic converters; environmental regulations affecting the use of high-sulfur fuels and the costs of expanding domestic mining and drilling; and federal land-use and leasing policies. The costs of shifts in such policies are largely the environmental or other benefits foregone. Such policies can be of some short-term usefulness, especially in increasing substitution possibilities, but more importantly, they have long-term effects on substitution possibilities and import dependence.

General Subsidies to Domestic Production

The benefits of domestic-production subsidies accrue from lower import dependence and greater domestic capacity in the event of a disruption. The direct economic costs consist of the extra costs of domestic production, over and above the prerestriction cost of imports, that producers are induced to incur. Subsidies will be more effective in increasing domestic capacity and production if there is a greater long-run elasticity of domestic supply, that is, a greater domestic resource base and more favorable technological and economic possibilities for resource exploitation. As with other measures increasing domestic production, subsidies can speed the exhaustion of the resource base and decrease future supply security. Offsetting this negative effect, subsidies could directly or indirectly induce exploration, discovery, and expansion of the known resource base. As the benefits of subsidies come from increases in capacity, they are effective largely as a long-term measure, implemented well in advance of any supply disruption

Subsidies redistribute income from the general taxpayer to domestic producers of the commodity and to users who benefit from lower prices. In doing so, they create dependent domestic interests. The government budget of course suffers. Distributive effects may be lessened, at the cost of administrative difficulties, by granting subsidies to particular domestic groups. The distributive effects also may be lowered by varying the subsidy inversely with the world price, so as to guarantee domestic producers an amount sufficient to call forth the desired increase in output; such a mechanism could also decrease the price uncertainty faced by domestic producers and so yield greater increases in domestic capacity than a fixed per-unit subsidy.

Subsidies can also have a deterrent effect on foreign producers by lowering the cost to domestic users of shifting to domestic resources, and so placing a ceiling on the prices other countries could charge without losing the U.S. market. Such a threat would be credible only if producers outside the United States believed it would actually be implemented, and if the domestic resource base could plausibly supply U.S. consumption. Actually carrying out such a threat would, in the absence of a restriction, impose a substantial burden of increased production costs.

Import Tariffs

By raising the domestic price of imports, tariffs lessen both import dependence and restriction impacts by decreasing overall use of the commodity and increasing domestic production and capacity. These adjustments impose costs similar to those imposed by restrictions: increased production costs and losses due to decreased consumption and costly user substitutions. Domestic producers acquire a vested interest in sustained high prices. Tariffs will be more effective in lowering restriction impacts if there are greater price elasticities (flexibility) of domestic supply and demand, and a larger gap between the short- and long-run consumption and production possibilities. As with subsidies, tariffs must be implemented well in advance of restrictions to yield major benefits. They will be relatively unattractive when production and consumption are inflexible and do not respond to price changes.

As with subsidies, tariffs redistribute income, in this case, from users to domestic producers in the form of higher prices and to the treasury in the form of tariff revenues. For a given import reduction, a tariff imposes lower costs on the economy than a subsidy does: the burden of adjustment is shared among producers and users. Under a subsidy, the full amount of import reduction must be made up by domestic producers; low-cost means of restricting consumption will not be adopted, as users will have no incentives to do so.

Tariffs may be used as deterrents, constraining producer-country power by raising import prices toward the cost of domestic alternatives. If effective in forcing lower world prices, tariffs allow the federal government to capture some of the potential monopoly profits of producing countries. As with subsidies, such uses of tariffs would be costly to the United States and might provoke some producing countries to extreme actions, such as embargoes or expropriation of U.S. properties. In addition, user groups would object to the tariff-induced escalation of prices.

In the context of planning for possible supply crises, tariffs have both benefits and costs: the United States can gain if import prices fall (that is, if a restriction ends), but domestic producers and users remain subject to price and demand uncertainty, and so may be deterred from investing in additional capacity or from implementing processes or technologies that would increase the ease of substitution. As noted above, to yield benefits by increasing domestic capacity and decreasing consumption, tariffs must be implemented well in advance of a restriction. Thus, the costs of a tariff in terms of distorted domestic resource allocations and the redistribution of income from users to producers and to the federal treasury will weigh heavily in any evaluation of policies for future crises. Under a fixed tariff, the price faced by domestic producers and users will vary directly with the world price.

Import Quotas

Quotas have the same basic effects as tariffs: decreased use of the commodity in question, increased domestic production, and redistribution to domestic producers due to higher domestic prices. Quotas, in the same manner as tariffs, impose costs by denying access to lower-cost imports, in the absence of a restriction, and by inducing higher-cost domestic production and costly reductions in consumption.

While tariffs necessarily redistribute income from users to the U.S. Treasury, the distributive effects of quotas depend on the method of allocating import rights. Producers gain if rights are granted to them without cost, but the government can gain, as with a tariff system, by selling import rights at auction or otherwise charging for their use.

Quotas could be used as a deterrent or threat in dealing with restricting countries, with the same risks of embargo or expropriation as with tariffs. Rights to sell under a quota system could be auctioned under sealed bids and be used to provoke competition among producing countries, especially if excess capacity is likely to result from restrictions and the resulting reduced demand.

Unlike tariffs, quotas do not allow the economy to benefit from declines in import prices, as when a restriction ends. However, they do lower the price uncertainty and risk faced by domestic producers and users, and provide producers with a more assured level of demand. On the other hand, under quotas, unexpected increases in demand cannot be automatically met by increased imports.

Government Stockpiling

The release of a government stockpile can lower the short-run costs of adjustment to a restriction, allow a more orderly and less damaging transition to alternative sources of supply, and lower resource transfer to foreign producers. Deterrence could result from the mere presence of a substantial government stockpile, combined with the belief among producing countries that the stockpile might be used in the event of a nonmilitary crisis.

The costs of stockpiling include the interest charges on capital tied up in stockpiled resources and storage facilities, depreciation, and administrative costs. In addition, government stockpiles and expected releases may reduce private stockpiling and other forms of private contingency planning.

Stockpiling is potentially a highly flexible policy, requiring a shorter lead time than tariffs or subsidies. However, to be effective, it also must be implemented in advance of restrictions. The main risk in rapidly building up a stockpile is in driving up the acquisition costs, and hence reducing the potential net economic benefits.

In the event of a restriction, government-stockpile releases can lower domestic prices, redistributing income by moving it away from private stockpilers and toward other users. If the government sells stock at market prices, it may, in effect, capture some of the income which would otherwise have accrued to producing countries. We found some stockpiling to be justified in all of the markets studied.

Subsidies and Other Encouragement of Technological Change

Encouragement of technological change can yield the following benefits via all of the mechanisms discussed above: increasing domestic demand and supply flexibility, reducing import dependence, speeding adjustment to price changes, and, through all of these effects, constraining producer-country power. The explicit mechanisms of encouragement can range widely, including:

1. operating subsidies to users of particular technologies;
2. capital subsidies to users of particular technologies;
3. price supports or purchase guarantees for users of particular technologies;
4. government subsidies for research and development;
5. government performance of research and development;
6. government ownership and operation of technically advanced facilities, on a pilot, ongoing commercial, or standby basis;
7. tax incentives for implementing processes which minimize the use of critical imported materials; and
8. government-sponsored assembling and dissemination of technical information.

Whatever the mechanism utilized, the economic costs in the absence of a supply restriction will be basically the additional value of resources used, over and above the cost of imports. In addition, especially for projects of reasonably near-term commercial viability, government support of technical change, and particularly government performance of research and development, may slow private development efforts.

Most types of technological adjustments, and hence policies to encourage technological change, are more suited to long-term anticipation of restrictions than to short-term responses or contingency planning. Technological policies may be complementary to shorter-term measures, such as stockpiling. For example, if long-lasting restrictions appear likely, early research and development efforts could ensure the availability of technology for substitution and transition to the use of domestic resources. Stockpile releases could then mitigate the short-term impact of restrictions.

Policy makers should not overlook the technological impacts of nontechnological policies. Tariffs or quotas that drive up materials prices encourage a search for demand and supply alternatives; subsidies for domestic production or other measures tending to lower prices may discourage such searches. Private-sector expectations of substantial government-stockpile releases in the event of a crisis may lead not only to lower private stocks, but also to decreased interest in the development of technological alternatives.

Price Controls and Allocation Plans

Because price controls and allocation plans are unlikely to decrease the real economic costs of a restriction, we did not analyze them in detail. The economic costs of a restriction consist not of the higher prices for various goods and services, but of the resource transfers to foreign producers, increased domestic production costs, and efficiency losses to users. Suppressing higher prices by controls will further exacerbate shortages, and prevent allocation of scarce supplies to their highest-value users.

Controls, combined with government plans to allocate existing privately held stocks, can lower private incentives for contingency planning and result in lowered efficiency. Price controls in combination with allocation plans can have desirable political effects. Visibly inflationary impacts may be lessened, but the gains of producers who, through accident or design, prepared in advance for the supply restriction will be lessened. Of course, if such measures are anticipated prior to a disruption, then incentives for private contingency planning are lowered.

Policy Flexibility and Vested Interests

Our analysis points to some underlying difficulties related to the distributive impacts of both restrictions and policy tools.

Policies raising domestic prices and production can create groups with vested interests in maintaining high prices. Inefficient producers sustained by tariffs, quotas, or subsidies may urge the maintenance of such policies even during supply restrictions. Tariff, quota, and subsidy programs can confer very large benefits on current domestic producers; political acceptability may have to be purchased at the cost of excess-profit levies or other forms of taxation. Similarly, the political acceptability of likely higher private profits and capital gains during a restriction may have to be purchased by some forms of taxation, with part of the cost being the impact on private incentives. The petroleum experience suggests that fuel taxation is most unlikely.

While domestic producers and resource owners gain from some policies, users, who will lose from high-price policies, will constitute a powerful force against many potentially useful actions. For example, while a tariff may impose small costs on the nation as a whole, users will see tariff payments as a cost and notice no definite or direct gain to themselves. Thus, policy makers may find implementing subsidies easier than tariffs or quotas, even though the balance of economic merit may in some cases be more on the side of tariffs or quotas. On the other hand, subsidies require explicit financing by the U.S. treasury, while tariffs and quotas are, in effect, financed by transfers from users to producers.

The implementation of stockpiling poses other problems. First, there may be pressure to build up the stockpile from domestic sources at premium prices, creating a vested interest in large stockpiles and high domestic prices. Such strategies could lead to inefficient development and exploitation of domestic resources. Second, in the event of a supply restriction, the stockpiling authorities may be pressured to allocate available stocks on a nonprice basis. Sales at the higher, postrestriction price may be condemned as government profiteering. It is possible that such restrictions will interfere with the efficient utilization of stocks. Policy makers may feel, on the other hand, that preservation of existing market structures and competitive positions are more important than maximum efficiency and so choose to allocate stockpile releases on a nonprice basis.

Policy Choice and the Form and Duration of Restrictions

Policy makers will never be certain of the opening date, duration, or severity of a restriction. Any policy which encourages producers or consumers to take actions not commercially viable in the absence of a restriction will impose net costs, if we set aside possible deterrence effects. Such policies yield benefits only if a restriction occurs. As a first approximation, the benefits are realized only for the period of the restriction. On the other hand, the most effective policy programs must generally be put in place in advance of a restriction, and hence costs are incurred over the whole planning period. The benefits of reduced restriction costs therefore must be sufficient, when weighted by the restriction probability, to outweigh these costs.

Severe producer-country supply restrictions have been very rare; OPEC is the most striking example. Large policy investments can be justified only if they can produce exceptionally large benefits when in fact a restriction occurs, when restrictions appear quite likely to occur in the near future, or when policy makers place a very high value on insurance against unlikely events.

Because of its flexibility and potential for rapid supply response, stockpiling is particularly useful if embargoes or short-lived and severe restrictions appear likely. Yet if longer-lived and less severe restrictions appear likely, tariffs, subsidies, and other policies yielding a permanent reduction in import dependence may be useful. This merit in such circumstances is purchased with sustained increases in costs to consumers, the creation of vested interests, and domestic income redistribution.

OPTIMAL POLICY MODELS

Important tools in analyzing the costs, benefits, and transfer effects of various policy options have been the computerized optimal policy models, developed by the authors, that simultaneously (or separately) evaluate various nontechnological and technological policies. Information required for the models includes: (linear) approximations to supply and demand functions; probabilities of supply interruptions; the market discount rate; stockpile carrying costs; and speeds of adjustment in domestic supply and demand in response to price increases or embargoes. The outputs of the model include: expected costs of supply interruptions; economically efficient production subsidies and tariffs (or import quotas); transfers among consumers, producers, and the U.S. Treasury; and the contingency benefits of technological policies. The Appendix to this volume describes the type of model used in many of this book's analyses.

The most important feature of the optimal policy model is its simultaneous evaluation of policies. The effects of technological and nontechnological policies are evaluated on a consistent and interactive basis, rather than being studied in isolation. For example, a technological policy that reduces consumption could be extremely valuable if no other policies were pursued, but the simultaneous implementation of an economically efficient stockpile program might reduce the benefits of that policy considerably.

3
RISK ASSESSMENTS IN PARTICULAR MARKETS

Our investigation of the seven selected commodity markets yielded substantial insights into both the general problem of contingency planning for supply crises and the proper role of technologically oriented policies in that planning. These investigations firmly support the principle that risk assessment should be based on a detailed analysis of the relevant markets, and that in situations of substantial risk, contingency policies must be based on a careful assessment of demand and supply alternatives.

As the purposes of this book are to present policy conclusions and compare appropriate instruments in differing market situations, we do not present detailed summaries of the analyses. Rather, we concentrate on the basic factual underpinnings of the conclusions and attempt to generalize from the seven case studies. Figure 3.1 shows the factors taken into account in these analyses.

THE LIKELIHOOD AND FORM OF DISRUPTIONS

The task of risk assessment must proceed in several stages, as shown in Figure 3.2. First, present and likely future import dependence must be appraised; if dependence is low, the United States faces no substantial threat. The next step is to investigate the actual concentration of production and its potential evolution in the future. If no single country has, or appears likely to command, a substantial market share, a damaging unilateral restriction is not possible; if no small group of countries controls a substantial share or production, then concerted restrictions are unlikely. Reserves and resources assessments are needed to identify likely long-term trends in concentration as well as potential supply sources.

FIGURE 3.1

Elements of World Materials Markets

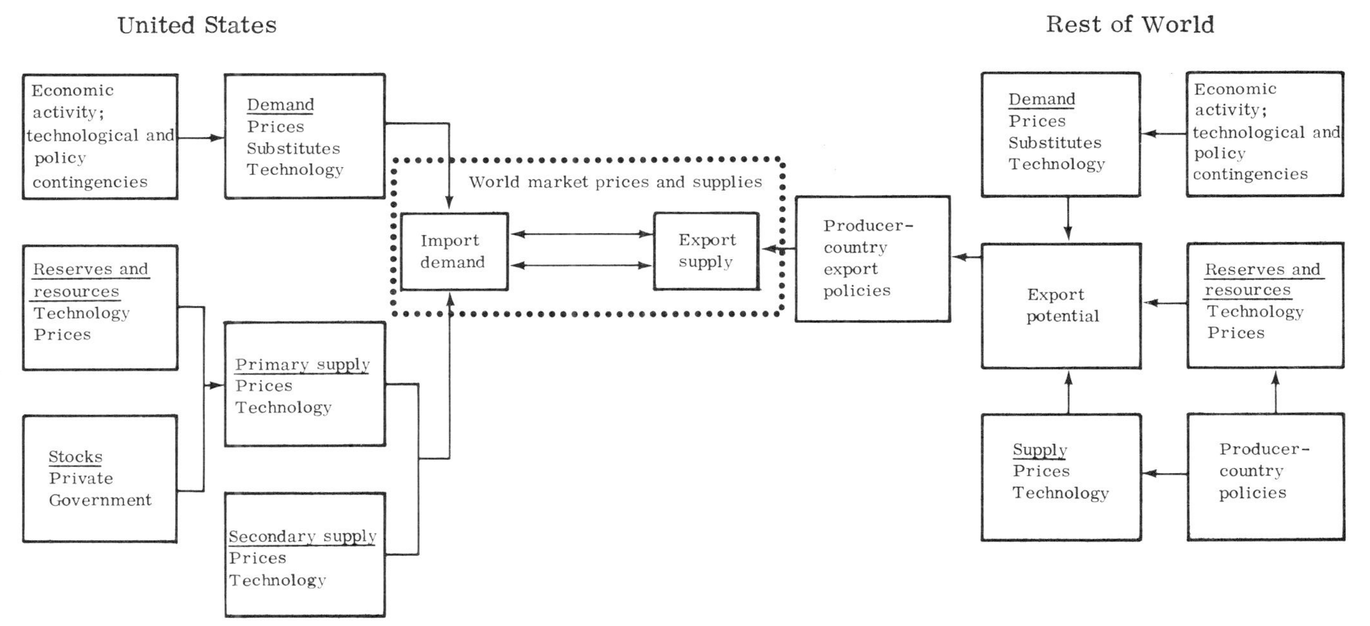

Source: Charles River Associates, Policy Implications of Producer Country Supply Restrictions: Overview and Summary (Cambridge, Mass.: CRA, 1976), p. 24.

FIGURE 3.2

Elements Required for Risk Assessment

- Import dependence: present and future
- Concentration of world production, reserves, and resources
- Producer-country intentions
- Potential profits from restrictions

Source: Charles River Associates, Policy Implications of Producer Country Supply Restrictions: Overview and Summary (Cambridge, Mass.: CRA, 1976), p. 25.

Producer-country intentions may be inferred from their observed market behavior, from capacity-expansion plans, from the role of the relevant commodity in producer-country economies, and from development and other plans. Such assessments also depend on evaluation of the potential economic gains from restrictions.

The information requirements for risk assessment overlap with those for impact evaluation. The critical determinants of the potential profitability of restrictions are the price flexibilities (elasticities) of demand and of supply outside the restricting country or group of countries. Low flexibilities of demand and alternative supply have two critical implications for restricting countries. First, low flexibility means that producing countries can raise prices substantially without suffering large sales losses, and thus can increase their earnings. Second, particularly important for cartel actions, low flexibility of demand and alternative supply implies that only relatively small output restrictions would be required for profitable price increases. If only small cutbacks are required, disagreements as to the division of the cutbacks are less likely than with large restrictions, and cartel stability will be promoted. Low elasticities of demand and alternative supply also indicate large potential damage to consuming countries.

The factors considered in assessing the risks of supply disruptions include those discussed above, as well as factors peculiar to the individual markets. Demand-side substitution possibilities are referred to, but discussed in more detail in the following section on impact assessment. Markets are considered approximately in order of their degree of risk as weighted by possible disruption impact.

PETROLEUM

In the petroleum market, while external pressures and internal conflicts will surely take their toll, OPEC's collapse does not appear imminent, U.S. import dependence will continue to be substantial, and there is a high risk of future embargoes or sharp cutbacks to serve political ends or to reestablish cartel discipline. There is also, of course, a significant risk of supply problems caused by political or wartime disruptions.

The working assumption used in our analysis is that in the absence of major structural changes in the market during the forecast period OPEC would maintain oil prices in real terms at the levels reached after the 1973-74 crisis ($12.73 per barrel delivered in the United States in 1973 dollars). In fact, OPEC prices have risen above this level in real terms, but not by enough to affect our conclusions significantly. Forecasts of oil supply and demand indicate that OPEC will not be subjected to major outside competition in the near future. OPEC reserves and capacity are so large and production costs so low, relative to those in other petroleum-producing countries, that OPEC will continue to occupy a leading position in the world oil market well into the twenty-first century.

We found that cost differences among OPEC members are relatively small and unlikely to lead to substantial conflicts over pricing policy. Similarly, quality differences exist, but are too widely recognized to serve as the pretext for massively destabilizing competition.

Given the relative inflexibility of petroleum consumption and alternative supplies, including alternative energy sources, OPEC clearly has a substantial profit interest in maintaining high prices. The critical question is whether this common interest in high prices will overcome the divergences among the countries. OPEC is far from a monolithic and perfectly coordinated entity. During the 1973 embargo, the non-Arab members did not support the cutbacks, and even Arab support was not uniform, though all members endorsed the price increase.

Excess capacity varies widely over time and among OPEC members; as of 1976, it varied from only about 10 percent in the case of Iraq to about 67 percent for Libya. During much of 1975 Saudi Arabia's unused capacity alone accounted for over 40 percent of OPEC's idle capacity. If OPEC members become unwilling to bear the burden of excess capacity and forego the potential earnings that this capacity represents, then competition will break out and the price will decline. Saudi Arabia's apparent willingness to behave in a statesmanlike fashion and bear the burden of excess capacity is a major stabilizing force.

The OPEC members also differ substantially in their desires to increase production, as indicated by ratios of reserves to production (an indicator of a country's desires to exploit its resources). A high ratio of reserves to production can indicate a relatively low cost of developing new reserves and the substantial ability to increase production. Reserves-to-production ratios range widely among OPEC members, from about 13 for Venezuela to almost 70 for Kuwait (ignoring Ecuador, a very small producer). Except for Algeria, these ratios are higher for Arab than non-Arab OPEC members, which suggests that in the long run, the Arab members may desire relatively higher production and hence lower prices. Political factors may offset this tendency, as indicated by recent events in Iran.

We also found significant differences among OPEC members in capital formation as a share of gross domestic product, with figures ranging from about 10 percent for Kuwait to 40 percent for Algeria. These differences indicate variation in the need for oil revenues to finance economic development.

Finally, we found substantial intra-OPEC variation in the ratio of foreign exchange reserves to imports, indicating a difference in need for current oil sales to finance imports. In 1974 the ratio ranged from 30 percent for Indonesia to 387 percent for Saudi Arabia. While Saudi Arabia and many of the other Arab countries have high ratios of reserves to production, and hence may wish to expand output, their current need for oil revenues is low. As a result, they may not feel compelled to compete and undermine stability.

Marketing arrangements may also play a role in OPEC's future stability. At present, the distribution of output among countries is determined by decisions of the multinational oil companies, who act as a buffer between member countries and market forces. Whether removing or weakening this buffer (such as by oil-import quota auctions, discussed below) would substantially increase actual competition is difficult to appraise in the abstract. The history of past cartels does indicate that increased competition, or cheating from within, is perhaps the most frequent cause of cartel failure.

OPEC does not have strong political cohesiveness. While the Arab members are united in their stand against Israel, Algeria, Iraq, and Libya are anti-West and socialist; the other Arab oil countries are traditionalist and pro-West. Iran, though in the Middle East and Moslem, is not an Arab nation and has, until recently, remained outside the conflict. Iran's stance is unclear following the recent revolution and severing of relations with Israel. It is also unclear whether Iran's Islamic revolution will spread.

We thus find a mixed risk situation in the world energy market. OPEC has, thus far, proven to be a viable organization and has substantially elevated prices. Major alternative supply sources other than the North Sea and Alaska, which are accounted for in our analysis, are unlikely to emerge before 1990. Our analysis suggests that long-run profit-maximizing prices for OPEC are near the $12.73 level (in 1973 dollars) which was being maintained in 1974. However, this does not preclude oil prices from increasing faster than the rate of inflation, and politically motivated embargoes remain a distinct possibility. Supply interruptions caused by noneconomic emergencies in the Middle East are possible and could have far-reaching effects.

As indicated above, it is also possible that competition may break out and OPEC may collapse; such tendencies could be increased by consumer-country policies increasing supplies or lessening consumption. The discovery of major new petroleum provinces or unexpectedly rapid development of substitute energy sources could also undermine cartel stability.

Given these uncertainties, we consider four scenarios for the world petroleum market:

I. OPEC collapses in 1980 and the price of crude oil falls to $4.00 per barrel (in 1973 prices).
II. OPEC remains cohesive indefinitely and supports a price of $12.73 per barrel (in 1973 prices).
III. In 1980 an embargo reduces U.S. imports and raises the import price to $31.00 per barrel (in 1973 prices).
IV. An embargo occurs in 1990, reducing U.S. imports and raising the import price to $31.00 per barrel (in 1973 prices).

Scenario II seems most useful as a base case, and Scenarios III and IV represent disruptions, much more severe than occurred in 1973-74. If "disaster" scenarios (such as total loss of Persian Gulf supplies) are considered possible, much more draconian contingency measures than presented here for oil would be appropriate. Then recourses would still include substantial stockpiling, conservation incentives, and specific technology policies (which would have to be evaluated on a case-by-case basis). In all cases U.S. energy prices should be allowed to rise at least to world level, if not higher.

ALUMINUM/BAUXITE

The United States imports close to 90 percent of the bauxite it uses, including bauxite contained in the intermediate product alumina, and on the order of 10 percent of the aluminum metal used.

If present market conditions persist, this import dependence is likely to continue.

The United States and the rest of the non-Communist world face a producer association in the bauxite market—the International Bauxite Association (IBA). This association, composed of the world's major bauxite producers, led by Jamaica and others, includes Australia. Virtually all of the bauxite and aluminum presently imported into the United States comes from countries belonging to the IBA. Association members now account for about 80 percent of non-Communist production and 68 percent of the world reserves.

Jamaica, the dominant member of the IBA, greatly increased its taxes on bauxite, from an average of approximately $1.80 per ton in 1973 to about $15.00 per ton by 1975; excluding Australia, most other IBA countries have followed suit, raising taxes approximately to the Jamaican level. Brazil and the Cameroons are the major non-IBA reserve holders, with 37 and 11 years, respectively, of the world supply at the 1976 consumption rate.

Our detailed study of bauxite reserves, costs, and taxes disclosed several constraints on IBA power and indicated the most likely scenarios. The current tax structure is not stable because the aluminum companies have a substantial long-run incentive to increase the share of bauxite obtained from IBA members such as Australia, and possibly Ghana and Guinea, thereby decreasing dependence on traditional Caribbean sources. In addition, currently there are substantial incentives for the companies to increase production in Brazil and the Cameroons, which are not IBA members. The implication is that, unless Jamaica and other Caribbean countries are willing to tolerate continuing declines in their market shares, taxes must adjust so that delivered costs are roughly equalized in such major markets as the United States. This equalization could occur through decreases in the Jamaican and other Caribbean tax levels. In addition, Australia has reserves sufficient to allow substantial capacity expansion, and, relative to known reserves, current or planned production rates in Brazil and the Cameroons are much lower than is the case in the present major producing countries. Thus, any IBA strategy will have to allow for Australia, Brazil, and the Cameroons to expand their market shares. If the IBA successfully accommodates the expansion of Brazil and the Cameroons, the difficulties of rapid and secret output expansion, and the absence of other substantial potential competitors, may allow the IBA to maintain its profitable hold on the world market for some time.

Given the position of the IBA, we tried to determine its most likely policies over the long run as well as the potential for embargoes or other extreme short-run output or pricing actions. The possibility of short-run monopoly behavior is important since alumina plants are specialized to use particular types of bauxite; this limitation gives individual countries, such as Jamaica, great short-run power. However, short-run monopoly pricing or embargoes are very unlikely. While they could be damaging to the United States, as discussed below, the Caribbean countries, in particular, have a locational advantage in selling to the United States; bauxite earnings are substantial in relation to national incomes in these countries and account for large proportions of foreign exchange and government revenues. Short-run monopoly actions or embargoes could cause a combination of governmental and market reactions that could result in their permanent loss of the U.S. market.

If this analysis is correct, the principal issue is the long-run pricing policy to be followed by the IBA. The constraint on this policy comes from alternative sources of supply of aluminum, such as shifts among IBA bauxite suppliers, shifts to non-IBA suppliers, and, ultimately, shifts, at high prices, to nonbauxitic ores. The demand for aluminum does have significant elasticity in the long run, so that large increases in the price of aluminum would, over time, induce substitution of copper, plastics, steels, and other materials. However, at present tax levels the total cost of bauxite is only 12 percent of the cost of refined aluminum; excluding IBA taxes, the bauxite cost is only about 6 percent of the cost of refined aluminum. Thus, only extremely large increases in bauxite taxes would cause a noticeable decrease in the total demand for aluminum, whether from IBA, non-IBA, bauxitic, or nonbauxitic sources.

We consider two major scenarios. In the first, Australia and Brazil pursue a moderate tax policy of $3 to $4 per ton, in 1976 dollars; competition would then force the other producing countries to lower taxes to match these levels (adjusted for transportation and other cost differentials).

In the second case, we assume that Australia, Brazil, and the Cameroons match IBA prices (with a $2 per ton differential for Brazil and the Cameroons). In this case, Jamaican taxes rise from $15 to $20 per ton in 1976 dollars, remaining there until 1985, by which time such new producers as Brazil and the Cameroons may have expanded production substantially; after 1985, Jamaican taxes will rise further, toward the limit set by domestic nonbauxitic ores. We also consider the effects of two unlikely, but potentially more damaging, possibilities: an IBA embargo and short-run monopoly pricing by the IBA.

CHROMITE

Given the importance of the Republic of South Africa and the Soviet Union in the long-run success of a formal cartel in the chromite market, explicit OPEC-type collusion is highly unlikely. The United States is almost entirely dependent, except for stockpile releases, on chromite imports, and given the potential for instability of supplies from southern Africa (including Rhodesia) and the Soviet Union, the major risk apparently is that of interruptions in supply from one or more major suppliers, very possibly followed by collusion among the remaining suppliers. Major producing countries other than the Soviet Union, South Africa, and Rhodesia include Albania, Turkey, and the Philippines. While there are a number of minor suppliers, their expansion possibilities appear limited, except possibly for India and Brazil.

We judge explicit collusion to be unlikely for political reasons, and administering a chromite cartel would be difficult. The ore is highly heterogeneous, and demand depends on capital and durable-goods production, which fluctuate substantially. Reserve and production data indicate that market concentration may rise in the future, as the Turkish market share declines and, as the available information suggests, Soviet exports decline; the long-run prospect is for increasing U.S. dependence on southern African production.

While the current situation cannot be termed a crisis, it appears that recently the Soviet Union and Turkey have at least tacitly cooperated in the chromite market; the impact of this could be increased if Rhodesian supplies become unavailable. The very large reserves of ore in southern Africa indicate that South African and Rhodesian cooperation would be essential to the lasting success of restrictions in the chromite market.

Recent events in the chromite market demonstrate that both the impact and risks of disruptions depend on technological developments. Until recently, the major metallurgical uses of chromium generally required high-grade metallurgical chromite of types available principally from Rhodesia, the Soviet Union, and Turkey. The potential market power of these countries has been substantially reduced by two technological developments: solid state reduction of chromite (SRC) and argon-oxygen decarburization (AOD). Together, these developments have greatly increased the substitutability of the South African high-iron (formerly chemical-grade) ore and the high-chromium (metallurgical-grade) ores for production of ferrochromium, the standard intermediate product for producing stainless and other alloy steels. Such a development is of particular significance while the Soviet Union remains a major supplier of high-chromium chromite.

While there have been no significant unilateral or concerted producer-country actions in the chromite market, the effects of the UN embargo on Rhodesian chromium indicate both some of the difficulties which producing countries may experience in attempting policies of selective embargo or discriminatory price increases, as well as the impact on consumers of the withdrawal of a major supplier. Throughout the embargo period, there was evidence that Rhodesian chromite and ferrochromium continued to enter the world market, indicating the difficulty of enforcing selective embargoes or policing agreed-upon export restrictions among cartel members. The UN sanctions appeared to have little impact on U.S. consumption, though private stocks did decline and deliveries from the General Services Administration (GSA) stockpile cushioned the impact. Major domestic effects of the embargo on U.S. imports appear to have been the shifts from Rhodesian to Soviet chromite, and from Rhodesian chromite to Rhodesian-based ferrochromium.

Given the uncertainties surrounding future developments in the chromite market and the potentially severe impacts of disruptions, we report analyses of two scenarios of partial embargoes: the first reduced U.S. imports by 15 percent, and the second, by 26 percent. The second and more severe scenario corresponds roughly to a withdrawal of the Soviet Union, or to a South African withdrawal, with a partial offset provided by other supplying countries, such as Turkey and India. The scenarios analyzed thus bracket a range of reasonable possibilities.

MANGANESE

The United States depends on imports for all of its primary supply of high-grade manganese, an essential ingredient in steelmaking. Production is concentrated, with the top six countries—the Soviet Union, South Africa, Brazil, Gabon, Australia, and India—contributing approximately 90 percent. In addition, a number of other countries show substantial production and/or reserves, and manganese resources in undersea nodules are vast.

South Africa, which accounts for about 31 percent of non-Communist-world supply, contains by far the largest reserves in the non-Communist world. South Africa's resources appear to be enormous, but over time its production costs promise to rise as a result of increasing depth and declining grade. Because supplies from other countries are either declining or unlikely to increase significantly without substantially higher prices, it is likely that the share of world supply accounted for by South Africa will rise over time.

Australia, which accounts for 12 percent of non-Communist supply, is an important moderating factor in the market. Australia's reserves are sufficient to sustain its current output rate for decades, and it appears that it could increase capacity in response to substantially higher prices.

The other major current suppliers of manganese—Brazil, Gabon, and India—have market shares similar to Australia's, but are significantly limited as alternative suppliers, either in the short or long run. Brazil's highest-grade deposits may be largely depleted by the mid-1980s, although other deposits could come into use. Brazil has shown little tendency to collude in other markets, and in fact has refused to join the IBA or to tighten its bauxite-export controls.

Gabon differs from the other major producers in that its manganese exports account for a large share of its foreign exchange earnings. This indicates that Gabon might anticipate significant gains from large price increases. While its short-run expansion is constrained by transport bottlenecks, its reserves and resources could sustain future expansion.

India's past behavior in imposing manganese-export controls following Soviet withdrawal from the market indicates that it might cooperate with restrictions. In fact, export controls have contributed to reducing India's sales of high-grade ore. However, new high-grade deposits have been reported, and India may become a more important producer in the future.

The Soviet Union in recent years has exported principally to other Communist countries and has not been a major supplier to the non-Communist world; in fact, the Soviet Union has recently begun to import significant tonnages of manganese (from Gabon, among other sources). Reported reserves of manganese in the Soviet Union are large, but its reserves of high-grade manganese may approach depletion within our forecasting period (1990). The Soviet Union, however, could possibly become an exporter if prices were to rise sharply and to remain at high levels for a number of years. Assuming that the Soviet Union had the resources to support higher production rates, the temptation to increase exports would be substantial; if South Africa, Brazil, and Australia were the primary beneficiaries of a cartel action, the Soviets would be under little political pressure to restrict economically beneficial exports. The Soviets' actions in markets such as chromite and manganese suggest that they might be willing to collude on at least a tacit basis once they have attained a substantial market share. The Soviet Union did cut off manganese supplies to non-Communist markets before the Korean War. The main effect of that cutoff appears to have been the development of increased production in India, Brazil, and other areas.

It appears that the threat of explicit cartelization in the manganese market is moderated by the heterogeneity of the suppliers and by the importance of Australia and Brazil, who tend to be hostile to explicit cartels. On the other hand, tacit collusion among the biggest producers, constrained by the costs of production at less efficient deposits, could well become a recurring feature in the market, especially during periods of peak demand.

The major potential problems in the manganese market until 1990 arise from the possibility of damaging interruptions in supply from some sources, such as South Africa and Gabon. After 1990, unless new reserves are discovered, the Soviet Union changes its export policies, or manganese nodules are mined on a large scale, the non-Communist world will become increasingly dependent on South Africa for its manganese supplies. Monopoly pricing will become more likely, and the threat of damaging supply interruptions will therefore increase.

Given the uncertainties surrounding the possibility of a crisis in manganese supply, we consider two scenarios: a 15 percent and a 26 percent reduction in U.S. imports. The 15 percent disruption is itself large, but could be interpreted as a withdrawal of South Africa from the market, with some offsetting production increases in other countries, particularly Australia.

PLATINUM AND PALLADIUM

The United States imports essentially all of its supplies of new platinum and palladium, the most important of the platinum-group metals. There is virtually no chance that the United States will become self-sufficient in these metals, which are important to industry because of unique chemical and physical properties, particularly in catalytic and electric uses; platinum and palladium are also substitutes for other metals in some end uses.

Large quantities of platinum and palladium are produced in only three countries: South Africa, the Soviet Union, and Canada. The Soviet Union is the dominant palladium producer, with two-thirds of world production, and about 29 percent of platinum; South Africa is the dominant platinum producer, with approximately two-thirds of world production, and about 27 percent of palladium. Canada accounts for about 5 percent of world platinum and 6 percent of world palladium production. The United States and other countries account for small amounts, and their reserves and resources are very limited.

Thus, each metal is already at least partially monopolized. Soviet production and exports are, of course, centrally controlled and, presumably, directed to serve that nation's overall interest. Although South African production is split among several independent firms, one, Rustenburg Mines, is dominant, both in size and in terms of its historical role as the largest producer. Rustenburg typically holds very large inventories and acts as a price leader. The two largest firms together control 93 percent of South Africa's platinum-group production. South Africa's potential as a platinum monopolist is strengthened by the fact that in Canada and, importantly, the Soviet Union, platinum-group metals are obtained as by-products or coproducts of other metals production, particularly nickel and copper. This means that escalation of the platinum price is unlikely to substantially stimulate increased production, as might be the case in bauxite, chromite, copper, manganese, or petroleum. Similarly, the Soviet Union's position in the palladium market is strengthened by the much greater proportion of platinum in the South African ores and by its byproduct status in Canada.

It thus appears that the United States might face substantial risks of supply disruptions in these markets. While such disruptions certainly would be facilitated by the highly concentrated market structure, it does not appear that substantial future restrictions of exports would be in the producing countries' economic interest. The available evidence indicates that the South African and Soviet industries recognize their mutual interest in high prices and act in a tacitly collusive manner. Instances of aggressive price competition are difficult to find, and it appears that producers have already taken substantial advantage of their market positions. In many end uses the demand for platinum-group metals appears to be moderately price elastic in the long run, so large price increases could reduce total producer revenues sufficiently to reduce producer profits.

It is clear that the security of platinum and palladium supplies going to the United States is tenuous. The potential economic exposure is great, but less extreme than one might think. The threat of long-run cartelization is moot, as the producers are presumably already charging prices near monopoly levels.

The dangers of short-run disruptions are moderated by the existence of substantial stocks. In many end uses platinum and palladium are employed, in effect, as capital goods, and users hold large stocks in use. Thus, escalations in price, especially if perceived as temporary, can cause very substantial reductions in purchases, thereby substantially depressing producer revenues. Thus, price increases or attempted supply disruptions may be self-defeating. For example, in 1974 the Soviet Union attempted to nearly double the selling price of palladium within a few months. The results were large inventory liquidations by major consumers and months of depressed prices. Politically motivated embargoes or pricing actions are, of course, possible.

A long-run supply disruption caused by, say, protracted guerrilla warfare in South Africa and the disappearance of the Soviet Union from certain markets would cause severe problems.

However, a long-run disruption would also allow time for the development of substitutes in many cases.

In our analysis of the platinum and palladium markets, we analyze in detail the effects of cartels or embargoes of plausible duration, as well as the countervailing U.S. policies.

COBALT

The United States imports all of the primary cobalt it consumes, except for those amounts that are released from the GSA stockpile. The market is highly concentrated; Zaire, the leading producer and reserve holder, typically accounts for two-thirds of non-Communist production. The three African cobalt producers, Zaire, Zambia, and Morocco, together account for about 82 percent of non-Communist production. Developed producing countries, including Australia, Canada, Finland, and West Germany, together account for most of the remainder of world production, and these countries would be unlikely to join a cartel with the less developed African cobalt-producing nations. It appears that the current structure of the market would allow for an African cartel controlling over 80 percent of production. The potential market control of the African producers is further enhanced by the fact that, except for Morocco, cobalt is produced as a byproduct of other metal production, and hence is relatively insensitive to price changes. However, recovery rates in some countries might be significantly increased if prices rise a great deal.

As with platinum and palladium, the cobalt market is already effectively monopolized, and significant further restrictions may not be in the producers' long-run economic interests. Zaire, as the leading producer, acts as the price setter, and it sets the price in its own long-run profit interest.

While cartelization may not be of great concern in cobalt (because the market is already effectively monopolized), the consequences of supply disruptions could be severe. The politics and geography of Zaire suggest that it is not a secure source of supply, and recurring supply disruptions could occur even if Zaire had no conscious economic motivation to restrict supplies below the long-run monopoly level. The principal contingency we analyze, therefore, is cutoffs in supply from Zaire lasting one year or more.

COPPER

The United States is almost self-sufficient in copper, so there appears to be little significant threat to it from possible disruptions in the world copper market. Producing countries in the world copper market are, however, partially organized into an association, CIPEC, consisting originally of Chile, Peru, Zaire, and Zambia. In addition, forecasts using an econometric model of the world copper market (reported in the copper volume of the Charles River Associates [CRA] study performed for the Experimental Technology Incentives Program [ETIP][1]) indicate that, to maximize revenues, CIPEC will maintain a moderate price. In such a situation, U.S. dependence on imported copper may rise as high as 30 percent of consumption by 1990. This increase in import dependence will not be due to resource exhaustion as such, since U.S. copper reserves and resources are very large relative to current and expected consumption rates. Rather, import dependence is projected to rise because the cost of additional domestically mined copper is rising relative to the costs of copper produced abroad. Thus, the less effective CIPEC is as a cartel, the greater will be future U.S. import dependence on other producers, and vice versa.

CIPEC's market power is not great under current conditions. First, it controls only about 37 percent of primary copper production. Secondary production, a major factor in the market, constrains the short-run exercise of monopoly power: 53 percent of U.S. copper production and 33 percent of foreign production in 1975 came from scrap, and the supply of copper from scrap responds significantly in the short run to price changes.

In addition, the United States is a leading copper producer, accounting for 27 percent of world production, as compared with the 13 percent accounted for by Chile, the leading CIPEC producer. Canada, another major producer (12.7 percent) is not a CIPEC member and would be unlikely to join in concerted restrictions and could expand production and exports in response to higher prices. Thus, CIPEC does not include two of the largest producers, and its share of the market is so low that very substantial output restrictions would be required to have a major effect on the price.

Not only is CIPEC's short-run hold on the market tenuous, but the prospects for effective long-run control appear slim. Copper reserves and resources are widely dispersed; to control over half of world reserves, a cartel would have to include South America, Africa, Canada, and Oceania. Collusion among such a diverse group appears most unlikely. In addition, analysis of supply behavior using an econometric model of the market (see the copper volume of the CRA study performed for ETIP) indicates that over the long run (five years or more), supplies from Canada and other countries would be quite responsive to price. Planned expansion announced by producers in Papua New Guinea, the Philippines, the Republic of South Africa, Yugoslavia, and other countries also limits CIPEC power.

Announced expansion plans for the CIPEC countries as of 1976 indicated that they are not disposed substantially to restrict production. Realization of these plans implied a 9 percent growth rate for CIPEC output until approximately 1980, a moderate price policy, and some expansion of CIPEC's market share. Econometric simulations of the world copper market indicate that CIPEC's revenues would be maximized by growth of approximately 9 percent to 1980 and 6 percent thereafter. Substantial output restriction would reduce revenues, and thus more severe restrictions appear unlikely.

While CIPEC embargoes or shifts to a short-run monopoly-pricing policy are certainly possible, all the evidence indicates they are extremely unlikely under current conditions. Virtually all of Zambia's export earnings are from copper, as are three-quarters of Chile's and over 60 percent of Zaire's. Copper is Peru's single largest source of foreign exchange. In addition, copper production employs a significant part of the labor force in CIPEC countries; production cutbacks would result in politically unacceptable increases in domestic unemployment. Extreme pricing policies would induce reductions in consumption, increases in alternative supplies, and a decline in CIPEC revenues. For such actions to have any significant impact on the United States, all the CIPEC nations would have to cooperate, resisting temptations to compete and expand market shares.

These conclusions are consistent with the history of attempts to control the world copper market; organizations such as Copper Exporters, Inc. have controlled substantially larger shares of the market than has CIPEC, and have achieved little success. Though the threat appears very small, we analyzed the impacts of complete embargoes of copper imports under high, moderate, and low degrees of import dependence.

A more important issue in copper relates to the likelihood that U.S. import dependence will rise significantly in the future. It may therefore be desirable to adopt policies now to prevent significant security of supply problems in the future.

NOTE

1. Charles River Associates, Policy Implications of Producer Country Supply Restrictions: The World Copper Market (Cambridge, Mass.: CRA, 1976).

PART II

PETROLEUM AND BAUXITE: FORMAL CARTELS POSING SIGNIFICANT THREATS

Sheer market size and current dominance by effective cartels distinguish bauxite and petroleum from the other markets studied. Import dependence is less in petroleum than in bauxite, but the size of the petroleum market means that significant price escalation is a matter of grave concern. Both producer associations may well remain stable for a number of years; the risks do differ, however. While OPEC has demonstrated its willingness to impose embargoes, it seems quite unlikely that the IBA would do so. Key countries involved, particularly Jamaica, are extremely dependent on a continuing stream of bauxite earnings.

Energy use is pervasive in the economy, and increased petroleum prices are noticeable in individual family budgets. In contrast, bauxite accounts for a small share of total aluminum costs, and aluminum is neither as pervasive nor important as petroleum. Bauxite price escalations would not have the same kind of visible and traumatic impact as OPEC price hikes do.

The policy options for dealing with supply disruptions in the two markets also differ. Given the possible political interest of OPEC or individual countries in future embargoes, stockpiling is perhaps more critical in oil than in bauxite. New finds of bauxite outside the IBA countries may be more likely than major new oil discoveries, and U.S. resources of nonbauxitic ores are such that dependence on the IBA could be substantially reduced without major trauma. Most consumers would hardly notice the increased aluminum cost. In contrast to the development of synthetic petroleum, there is not a great deal of uncertainty surrounding new and alternative aluminum technologies. Issues of massive environmental impact do not arise with bauxite as they do with possible oil shale development or greatly increased coal mining.

OPEC's pricing policy and cohesiveness are better established than those of the IBA. If Australia ultimately does not cooperate by substantially raising bauxite taxes, Caribbean taxes are likely to fall in real terms in the long run. No such current tension threatens OPEC. This tension gives the United States leverage in the bauxite market. By giving favorable attention to Australia's larger political and economic concerns, the United States might encourage Australia to forego part of its narrow economic interests in the bauxite market. Although Saudi Arabia's desire to limit dependence on the Soviet Union encourages some degree of moderation in the oil market, U.S. leverage is probably less than in bauxite.

In bauxite, severe short-term disruptions seem to be less likely than for such markets as chromite and manganese, while in the long run, the IBA may escalate prices significantly and remain cohesive, and supply reductions caused by political disruptions are always possible. We examined the possibility of both short-run and long-run disruptions, and expended considerable effort examining the viability of long-run domestic supply alternatives. In petroleum the events of 1973-74 and the recent disruptions in Iran indicate that embargoes may recur. We therefore evaluated a wide range of alternatives in the context of possible future embargoes. Because OPEC will probably remain cohesive for some time, we appraised several options viable for the longer term—synthetic-fuels development, conversion of oil- and gas-fired electricity-generating plants to coal, and accelerated licensing of nuclear-power generating facilities.

Petroleum is by far the single largest item in world trade and U.S. imports. Bauxite, directly or as alumina, is the sixth largest item in U.S. imports; it follows iron ore, coal, grain, and phosphate rock. It is difficult to find other major materials markets of comparable absolute size in which the United States is sufficiently import dependent, in which producing countries have substantial commonality of interest, and in which market structure is concentrated enough to allow effective cartelization.

4
ENERGY MARKETS AND THE OPEC CARTEL: IMPACT AND POLICY ANALYSIS

The potential impacts of future energy crises dwarf the possible effects of crises in any other material market. Not only is the market huge, but all Americans are affected directly and substantially by changes in energy costs. Similarly, the potential gains from correct policy and, just as important, the costs of incorrect policy, are vast. The effects of even substantial and permanent escalations in price or supply interruptions in the other commodity markets studied would be almost negligible by comparison with small disruptions in energy markets.

The United States has already suffered the consequences of the 1973-74 crisis. The potential for future crises remains, and a vast array of policies has been proposed. Our purpose in analyzing the energy market was to compare a range of policies in a consistent framework, evaluating their economic efficiency, distribution effects, and other attributes. We proceeded in two steps, first attempting to draw messages from the energy crises of 1973 and 1974, and then evaluating them in the context of the four scenarios of possible future OPEC behavior presented in Chapter 3.

Though our choice of policies was necessarily restricted, those we have analyzed have been seriously proposed for at least initial implementation by 1990. Stockpiles, quotas, tariffs, and subsidies are the main policy measures that directly affect the domestic supply and demand balance for crude oil and petroleum products. All of these policies require the actual occurrence of an embargo or a cartel action to yield net economic benefits. In each embargo case, we evaluated the policies at their efficient levels—those yielding the maximum possible reduction in embargo costs, allowing for the costs of the policies.

We also evaluated selected technology-based measures designed to encourage alternatives to imported oil. We included the encouragement of synthetic-fuel technology, first because synthetic petroleum and natural gas can be substituted directly for petroleum products in major uses to a greater extent than can other energy alternatives, and also because the development of synthetic-fuels technology is at an appropriate stage to consider the issues of subsidizing new technology as a device for mitigating future energy crises. That is, synthetics technology is neither so advanced that the time for subsidized development has passed, nor at such an early stage of development that it cannot be considered a policy option for crises by 1990. In addition, we considered accelerated development of nuclear power and conversion of oil-fired generating plants to coal.

The policies examined present a range of merits, demerits, and risks. Subsidies of all sorts and stockpiles involve direct budgetary expenditures and place the burden of contingency planning on taxpayers. Tariffs, quotas, and subsidies can redistribute billions of dollars to domestic producing interests. Synthetics development, expansion of nuclear power, and, to a lesser extent, coal conversion entail environmental risks and costs. Synthetics development is faced with major cost and technological uncertainties.

Our main purpose in the present study is policy analysis covering possible future crises; we therefore first present our overall policy conclusions, then summarize the policy messages drawn from analysis of the energy crisis of 1973-74, and, finally, explore in more depth the merits of the individual policy tools and the analysis underlying our conclusions.

We did not examine in the study such policies as conservation (including both less consumption by final consumers and less use of fuels by industry) and stimulation of supply from such sources as solar energy, wood, and geothermal energy. Consideration of the vast array of possible policies was simply beyond the scope of our study, whose principal focus was to evaluate the economic merits of synthetic fuels production and other similar policies. Many analysts have made strong cases for alternative strategies. The merits of conservation options and stimulation of supplies from sources such as solar energy could be evaluated using the framework for analysis used in our study. Clearly, such options may be preferred to synthetic fuels if costs per barrel equivalent are less than those of synthetic fuels.

POSSIBLE FUTURE ENERGY CRISES: POLICY CONCLUSIONS

To decide which policy options or combinations of policies are most desirable, one must apply three kinds of judgments. First, one must judge the likelihood of occurrence for each of the scenarios under consideration, or for the class of outcomes that each scenario typifies: OPEC collapse (Scenario I), OPEC stability (Scenario II), an early severe embargo (Scenario III), or a late severe embargo (Scenario IV). The scenarios we examined do not bracket all possibilities. For example, future embargoes could be worse than we have assumed and future oil prices may rise faster than we have assumed. For both cases, more extreme contingency policies would be justified. Second, one must appropriately weight the public-interest value of mitigating the trauma of an embargo. Finally, subjective weights must be applied to the public-interest value of the various policy considerations, such as income redistribution, reversibility of policy effects, creation or perpetuation of special vested interests, and government fiscal effects.

We did not attempt to apply subjective weights to all of the factors that are essential to the formation of practical policy. Nevertheless, some general observations are in order. First, based on our analysis of the history of cartels and the factors bearing on OPEC's cohesiveness, we feel that outcomes of the sort represented by Scenarios I and II (OPEC collapse by 1980, and OPEC stability) are more likely to occur than a consciously motivated economic embargo. It should be noted that we do not explicitly consider supply reductions that might be caused by political or military disruptions. For an embargo of the severity represented by Scenario III (1980 one-year embargo) or Scenario IV (1990 one-year embargo), OPEC not only would have to remain intact as an effective monopolist for seven and 17 years, respectively, but it would also have to ascertain that an embargo would be in its self-interest, and show the collective will and cohesive strength to impose it effectively. Alternatively, one or more countries would have to cut their exports drastically, and others not act to offset this. The 1973-74 embargo ended after several months in part because it became increasingly difficult for OPEC to maintain discipline and cohesiveness over an extended period. The embargoes assumed in Scenarios III and IV last for a full year and are therefore much more severe than the 1973-74 embargo.

Second, the public attaches value to reducing the future trauma of an extreme economic event, such as a severe embargo, and to preserving continuity of essential supplies for national security purposes. However, none of the policy options considered would prevent a future embargo from becoming a crisis; they could only moderate it. Moreover, energy supply to meet national security needs can be assured much less expensively than through full-scale application of the policy options we consider; for example, special stocks could be held for national security purposes. Overall, concern over the effects of an embargo enhances the relative worth of stockpiles: per dollar of cost, stockpiles give the greatest short-run supply capability.

Finally, we can say little about the weight due the various policy factors that must be considered. The policies differ substantially with respect to these factors, and the factors may well determine actual policy choices. Conversion of generating facilities to coal and reduction of nuclear lead times fare well when examined from the point of view of these policy factors. They create few new policy-dependent vested interests, and involve minimal income redistribution and federal government expenditures. The major uncertainties and adverse policy effects concern possible environmental impacts. The other policies are more mixed, and their merits should be considered along with possible economic benefits under the various scenarios.

Tables 4.1 and 4.2 summarize benefit-cost information on the respective policy options. In view of the conditional nature of some of the underlying calculations, especially with respect to the cost of synthetic fuels, and in view of the importance of policy considerations not included in the benefit-cost calculations, the tables should not be used as a basis for a quantitative ranking of policies.

The major impact of future embargoes is clear. The first line of Table 4.1 shows that a 1980 embargo, equal in severity to the 1973-74 embargo but lasting for a full year, and raising the price of oil to $31 per barrel in 1973 dollars (Scenario III), would cost the United States $21 billion in 1973 dollars, not including macroeconomic consequences and political costs. By 1990 imports are expected to rise substantially, raising the cost of a similar embargo (Scenario IV) to over $62 billion. In the event of an OPEC collapse by 1980 (Scenario I), of the technology policies examined only coal conversion, which is commercially viable under either present or substantially lower oil prices, is economically beneficial. Compressing nuclear lead times would involve no substantial cost since construction schedules can be adjusted. The other policies impose net costs in this scenario, with a quota being particularly harmful, as it denies access to increased supplies of potentially low-priced oil. If OPEC is stable, the nuclear, coal, and oil shale policies would yield net benefits (see Table 4.2); under future embargoes,

TABLE 4.1

Net Benefits of Different Policy Options
under the Scenarios, 1975-90
(billions of 1973 dollars)

Policy	Scenario I	II	III	IV
No new policy	0	0	-20.66	-62.24
Stockpiling	-40.83 to -75.33	-34.13 to -68.63	0.83 to 7.46	-20.27 to -54.62
Tariff	-22.96	-7.05 to -11.47	13.70	15.11
Quota	-42.85	-7.05 to -11.47	13.70	15.11
Subsidies for domestic oil production	-4.52	-2.49 to -5.62	5.63	2.47
Conversion of oil-fired generation to coal				
Base case	1.68	11.20	15.31	12.31
Extended case	3.38	17.95	26.27	20.19
Reduction of nuclear lead times	0	positive, moderate magnitude	positive, same magnitude as in II	

Note: Benefits and costs are cumulated to 1990 (all scenarios), using an interest rate of 10 percent, compounded annually.

Scenario descriptions:

I: OPEC collapse by 1980.

II: OPEC stability through 1990.

III: 1980 embargo, equal in severity to 1973-74 embargo, but lasting for a full year.

IV: 1990 embargo, equal in severity to 1973-74 embargo, but lasting for a full year.

Source: Charles River Associates, Policy Implications of Producer Country Supply Restrictions: The World Energy Market (Cambridge, Mass.: CRA, 1976), p. 349.

TABLE 4.2

Maximum Gross Per-Barrel Benefits of Subsidies for Synthetic-Fuel Development (billions of 1973 dollars)

Fuel	Scenario I	II	III	IV
Oil shale	-4.55	4.18	4.18	4.81
Synthetic natural gas	-12.10	-3.37	-3.37	-2.74
Low-BTU gas	-4.34	4.39	4.39	5.02
Syncrude	-12.74	-4.01	-4.01	-3.38

Note: Cost figures used in this table to calculate gross benefits are weighted averages of the figures presented in chap. 6 of vol. 7 of the CRA report done for the ETIP. In evaluating these figures, the reader should keep in mind that it is assumed the target costs projected by the Synfuels Task Force in 1975 will be realized. Historically, there has been a tendency to revise such cost estimates upward, and the estimates make no substantial allowance for the opportunity cost of scarce water resources; this factor is particularly relevant to shale and to synthetics development using western coal. The estimates also do not make adequate allowance for the costs relating to opposition on environmental grounds. The latter includes costs of delay and retrofitting, and investments in additional protective measures not in the original plan to mitigate environmental effects. Recent experience with other energy technologies (for example, nuclear and fossil-fueled power plants and the Alaska pipeline) indicate these factors have contributed substantially to the general tendency of actual realized costs to be much greater than early estimates.

Scenario descriptions:

I: OPEC collapse by 1980.

II: OPEC stability through 1990.

III: 1980 embargo, equal in severity to 1973-74 embargo, but lasting for a full year.

IV: 1990 embargo, equal in severity to 1973-74 embargo, but lasting for a full year.

Source: Charles River Associates, Policy Implications of Producer Country Supply Restrictions: The World Energy Market (Cambridge, Mass.: CRA, 1976), p. 350.

all the policies, except for stockpiling, synthetic natural gas (SNG), and syncrude, can yield net economic benefits.

In most of the scenarios we analyzed, we did not explicitly consider indirect macroeconomic costs over and beyond the direct costs from higher oil prices. However, we did assume that oil prices would rise to market-clearing levels of approximately $31 per barrel in 1973 dollars, and the costs and benefits of the policies were evaluated at this price. By comparison, the average delivered price of imported oil in 1974 was about $12.50 per barrel, and the average price of oil to refiners was about $9.07 per barrel. It can be plausibly argued that many of the indirect macroeconomic costs during the 1973-74 crisis were caused by the distortions arising from price controls and direct allocations, as well as by the failure of the government to stimulate the economy to offset the OPEC price increase. It is to be hoped that the U.S. government will not repeat the same mistakes during a future emergency. In evaluating the policies at market-clearing prices, many, if not most, of the indirect effects can be assumed to be taken into account directly in the oil market.

As another way of looking at this, the average refinery cost of oil in 1974 was $9.07 per barrel, as noted above. One estimate of the indirect macroeconomic costs during 1974 is $24.51 per barrel. Using these figures, the total opportunity cost of oil during the embargo was $33.58 per barrel, which is quite close to the $31 estimate (in 1973 dollars) used in our scenarios.

The reader is reminded that the emergency scenarios we evaluated are extremely conservative ones and involve disruptions in oil supply that are much more protracted than actually occurred in 1973-74. Nevertheless, to provide upper-bound estimates of the benefits for one of the major policies, stockpiling, we evaluated the stockpiling scenarios using both the direct costs, based on a $31 embargo price, and an estimate of indirect costs of $24.51 per barrel.

As described in Table 4.1, when some level of a stockpile, tariff, quota, or subsidy could yield net benefits, we set its level at the efficient or optimal level in order to maximize net benefits (that is, reduced embargo costs), net of the costs of the stockpile. Where the efficient level was zero, we assumed levels optimal for the various embargo scenarios, which gives some notion of the possible costs of policy mistakes. Two levels of coal conversion are evaluated: the base case of coal conversions mandated by the Federal Energy Administration (FEA) under the Energy Supply and Environmental Coordination Act (ESECA), delayed three years to take account of conversion difficulties; and an extended, or the most optimistic, case, which assumes conversion of most plants on the original conversion list. The base case is optimistic but feasible;

the extended case is a reasonable upper bound. We estimate that in the base-case policy, coal will displace about 84 million barrels of imported oil in 1980 and 62 million barrels in 1990. The extended-case policy displaces 171 million barrels in 1980 and 126 million in 1990. The nuclear scenario assumes reduction of lead times to seven years, well below current lead times of ten to 11 years, and only 28 percent above the minimum time required for construction alone. This reduction is technically feasible but highly optimistic. The result is a reduction in the cost of nuclear generation sufficient to yield net benefits even under present conditions. The oil displacements are not large, as nuclear energy largely displaces coal base-load generation, not oil-fired plants.

When the economic benefits and costs and the policy considerations are taken together, the following pattern emerges. The conversion of oil-burning generating facilities to coal is a highly cost-effective measure that can have a substantial effect in mitigating the costs of future embargoes. However, environmental considerations limit the degree to which conversion can be effected beyond the limits now agreed to by the Environmental Protection Agency (EPA) as of 1977. Research and development to improve scrubber reliability could allow further expansion. Moreover, the benefits of conversion to coal decline with time as the stock of oil-burning plants grows older.

Reducing nuclear lead times appeared highly cost effective as a means of reducing energy cost. However, it can have little effect on oil imports and can mitigate an embargo only after 1985: nuclear plants generate base-load electricity and do not directly displace much oil, and minimal lead times for nuclear-plant construction are substantial.

The efficacy of stockpiles, tariffs, quotas, and subsidies depends on the weight given to the various judgmental factors discussed above. It is not clear that any of these policies is preferable to the benchmark option of doing nothing at all. Even using the calculated net benefits of each of these policies as the sole basis for evaluation, an embargo occurrence must be a fairly probable one to make the policies an attractive alternative. Moreover, using calculated net benefits of these policies without adjusting for conversion of generating facilities to coal or for other policy options of factors biases the analysis in favor of each policy in question. Coal conversion would reduce embargo benefits from each of the policies, and policy factors such as income redistribution or creation of vested interests generally have a negative rather than a positive influence on an analysis of these policies.

Tariffs, quotas, and subsidies fare somewhat better than stockpiling if we apply the same hypothetical embargo weights, but the policy factors for tariffs and quotas may make them less acceptable options to the public than stockpiling. From a narrow dollar benefit-cost standpoint, tariffs are preferable to quotas. If implemented well in advance, subsidies for domestic crude production could help provide a quick-response capability that would be helpful in the event of an embargo. However, subsidies may not fare as well as other policies on grounds of income redistribution and creation of vested interests, particularly vis-à-vis tariffs and quotas. Finally, the use of capital grants and other substantial subsidies for commercial-scale development of synthetics has adverse effects in terms of income redistribution, irreversibility, creation of vested interests dependent on continued subsidy, disincentives to efficiency, and a potentially open-ended commitment to large federal government outlays over a long period of time.

Our evaluation of the costs of synthetic fuels and of related policy options leads us to the following conclusions. First, there is an important policy difference between shale oil and low-Btu gas, on the one hand, and between high-Btu gas and syncrude on the other. The latter two technologies have such high projected costs before 1990 that it appears that no reasonable policy alternative could effect large-scale commercial production in time to mitigate the costs of such an embargo. High-Btu gas and syncrude offer longer-term promise, and policies encouraging research and development to reduce their future costs may be a good investment for the long-term future in energy supply. But large-scale capital subsidies or price supports high enough to stimulate commercialization by 1985 would be a costly policy.

Shale oil and low-Btu gas, however, are apparently much closer to commercial viability. Implementation of these technologies may be deterred by uncertainty over OPEC's pricing strategy and by the risk of OPEC collapse. There are also institutional obstacles involved; in the case of shale oil, there are the problems of water rights, the use of federally owned land, and opposition on environmental grounds. Resolving uncertainties over environmental, land, and water policies, and setting a modest support price at, say, 75 to 85 percent of the present OPEC price would probably provide a more stable and predictable environment for commercial development. If policy leaders would accompany such a policy with a straightforward disapproval of capital grants or other direct subsidies, any tendency to delay development in anticipation of large subsidies would be avoided. If the technologies of shale oil and low-Btu gas are potentially efficient supply sources by 1985, the private sector should assure development of these technologies under the policies set forth here. Heavy government subsidization would involve excessive public expenditure, disincentives to efficiency, and the perpetuation of a subsidy-dependent vested interest.

One reason for embarking on at least limited development of synthetics, as opposed to concentrating on nontechnological options that may appear to yield higher net benefits at present costs, is to establish and strengthen the credibility of our long-run alternatives to imported oil. The same holds true for coal conversion. Potentially fruitful areas of technological research in synthetics include: resolution of problems of in situ retorting that could lower water requirements, mining costs, and environmental damage for shale oil; investigation of improved methods for using domestic coals in the manufacture of synthetics; resolution of cost uncertainty by pilot plant operations. Such investigations might be done by government, with industry consultation; by industry, under government contract or subsidy; or in a variety of other ways. Coal-conversion prospects would be improved by research in a number of areas, including improved scrubber reliability, and methods for clean combustion of higher-sulfur coal, including coal cleaning and fluidized-bed combustion. A final critical technological area is the standardization of reactor design. In the present policy environment, compressed lead times may be achievable only through standardization. The federal government could continue its efforts in this area and encourage state power-plant siting authorities to adopt streamlined procedures taking account of this standardization. Nuclear lead-time acceleration seems attractive if evaluated strictly on cost-efficiency grounds; however, there would be only minimal contingency benefits.

THE ENERGY CRISIS OF 1973-74

Because the 1973-74 trauma is the most severe peacetime materials crisis that has been experienced by the United States, we investigated the nature of public and private anticipation of and response to it. We also evaluated policies which could have been adopted, had the crisis in fact been anticipated in 1969. This examination gives further insight into the merits of the various policy options.

The policies considered were, of course, not utilized during the 1973-74 energy crisis. The most significant result of our policy analysis is that it shows that the subsidization of synthetic fuels would have been a very inefficient policy choice and one of doubtful feasibility. Only oil shale and low-Btu SNG appear to have been developed enough to have been feasible policy tools at the time. Even if SNG, the most promising synthetic fuel, had achieved the optimistic cost level assumed in the analysis, it would have provided energy at a price only slightly below the embargo price. Because such cost estimates are not completely reliable, it is unlikely that the policy would have yielded net benefits. Oil shale fares worse than SNG by narrowly economic criteria and suffers further when the likely environmental consequences are factored in.

Stockpiling, import quotas, and tariffs show distinctly larger net benefits than do subsidies for domestic production. Though the calculated net benefits for quotas and tariffs are greater than for stockpiling, the latter may be preferable on income-redistribution grounds because of the large transfers from energy users to producers that a tariff or quota would imply. While stockpiling would have involved continuing costs and a substantial effect on the federal budget, it would not have created a substantial vested interest in higher prices or a policy-dependent class of inefficient producers.

Compression of nuclear lead times does not appear to have been a relevant policy for the crisis. During the 1969-73 period nuclear lead times were not significantly above feasible minima; standardized reactor designs were not fully available or accepted. Preservation of the coal-burning capability of dual-fuel (coal-oil) generating plants could have mitigated the crisis and yielded significant net benefits. It is possible that an early anticipatory coal-conversion policy, like that later mandated by ESECA, would also have yielded net benefits, while avoiding other negative policy effects. Uncertainty about scrubber technology and environmental impacts, however, would have been massive.

Economic Impacts of the Energy Crisis

The energy crisis and the subsequent higher prices had, and continue to have, sweeping effects on resource allocation, government policies, and, in the long run, life styles. Sizes of automobiles have changed, exploration for oil has increased, and the development of alternative energy sources has accelerated. Government policy debates have come to center on energy problems.

During the crisis, partially as a result of price controls, some users were unable to obtain oil at any price, just as others were unable to obtain oil-based products. We did not attempt to evaluate the benefits and costs of all of these adjustments and impacts. Rather, we concentrated on plausible estimates of the economic impact.

The effects of the embargo, combined with the OPEC production cutbacks and price increases, were to raise the average price of crude by a factor of over three and to reduce imports over the crisis period by an estimated 37 percent. In terms of higher oil costs and lost consumption, these effects imposed substantial direct costs of $14.5 million per day in December 1973 and, as oil prices soared, $29.4 million per day in March 1974. Over the crisis period the direct economic cost was about $2.7 billion, or $8 billion on an annual basis. For an average import shortfall of 1.9 million barrels per day, this amounts to a direct cost of about $14 per barrel. Using estimated demand responsiveness to take rough account of direct quantitative shortages or of the inability of some users to obtain products, the costs are higher: about $7.6 billion over the embargo period, or $22.8 billion on an annual basis for a per-barrel cost of over $39.

In addition, the oil embargo and price increases, by a combination of expectations and income-multiplier effects, may have reduced 1974 GNP by between $16 billion and $34 billion. To an unknown extent, these cost estimates, attributed to the oil embargo, are the effects of a developing recession.

What general message can be drawn from the impact of the energy crisis? In the case of oil, the relative short-run inflexibility of consumption and production of oil and other energy sources, both in the United States and the rest of the world, meant that the OPEC cutbacks could both sustain the more-than-tripled price and cause substantial damage. Stocks were low (for most products, less than two months' supply), and so could not serve as an effective buffer. Thus, it appears that if foreknowledge of the crisis had been possible, effective government contingency planning would have yielded substantial economic benefits.

AN EVALUATION OF THE MERITS OF INDIVIDUAL POLICY TOOLS

Basic Methods of the Policy Analysis

In the remainder of this chapter we describe in more detail the results of our analysis of both specific policies and impacts in the absence of policies. The method we used was to estimate the amount of imported oil that might be displaced by the various measures, and the cost of obtaining this import reduction. Gross policy benefits were obtained by valuing the imports that would be displaced during possible future embargoes; net policy benefits were obtained by subtracting the cost of obtaining this reduction. Where relevant, we evaluated policy impacts on income distribution among producers and consumers. For the embargo scenarios we evaluated the optimal or economically efficient levels of stockpiles, tariffs, quotas, and subsidies—the level of the instrument that would yield the maximum net benefits (that is, the reduction in embargo costs minus the policy costs). If one knew with certainty that the embargo would not occur, the optimal levels of such policies would be zero, since there would be no embargo costs to mitigate, and the policies themselves impose costs. Because OPEC's future behavior is uncertain, we assessed policy risks by evaluating policies optimal for future embargoes in the nonembargo situation.

It was not highly meaningful to evaluate efficient levels of the synthetics, coal-conversion, and nuclear policies: very large synthetics development could violate environmental or resource-availability constraints; nuclear lead-time compression is limited by time needed for construction and by public policy considerations of safety and complete review; and coal conversion is limited by environmental considerations, and by the simple fact that the amount of oil burned in power plants is not a major component of U.S. oil consumption. Conversion can spare no more oil than is in fact burned. Because of the uncertainty about costs and feasible scales of development, we evaluated synthetics on a per-barrel basis, determining whether they could yield benefits under the various scenarios. We evaluated two relatively optimistic, feasible coal-conversion scenarios. Nuclear acceleration was evaluated for estimated minimum realistic lead times and realistic amounts of oil displacement.

The basic parameters were obtained from the FEA's forecasts of U.S. production and consumption in the National Energy Outlook. Production and consumption responses to changed energy market conditions and policy instruments were based on available evidence, and on our own calculations based on recent experience. The scope of the project did not allow for the construction of full-scale models of energy markets. Rather, we used plausible parameter estimates to evaluate all policies in a consistent framework. This, and not forecasting or energy market modeling, was the focus of our study.

Formulation and evaluation of the technologically oriented policies—synthetic-fuels development, coal conversion, and compression of nuclear lead times—required appraisal of technological uncertainties and environmental risks. Our policy analysis reflects these appraisals.[1]

Impacts in the Absence of Specific Policies

Under Scenario I, the OPEC collapse, there are no future crises, and hence, no crisis costs that policies might mitigate. Rather, the economy reaps substantial benefits of $8.73 per barrel in real terms, for a total 1980 benefit, in terms of savings on imports, of about $57 billion; the total benefit is, of course, even greater due to the consumption gain and smaller high-cost domestic production. The benefit to the United States from the OPEC collapse would increase over time, and in total would be huge—in the hundreds of billions of dollars.

Obviously, any policy making OPEC's demise substantially more likely would yield huge savings. Policies decreasing the demand for OPEC oil or otherwise increasing divisiveness or competitive pressure may have this effect. However, a review of the combined efforts of all the policies we considered, including those that are hopelessly uneconomic, suggests that the decline in imports, if the United States adopted an immense and coordinated program of development of new technologies, and direct and indirect incentives to users to economize and to producers to increase supplies, would still be fairly small relative to total OPEC sales. This suggests that for such policies to have a significant chance of disrupting OPEC, they would have to be pursued in conjunction with equally severe policies by the other consuming nations. On balance, it seems likely that coordination on a sufficiently large scale is simply not feasible.

Under Scenario II (OPEC stability, and price stability in real terms), the economy continues to incur substantial costs equivalent to the benefits discussed for Scenario I. The $4.00 per-barrel cost of oil in Scenario I is taken as a base.

Under Scenarios III and IV, the economy incurs substantial costs. In the absence of government policies such as stockpiling or other measures to decrease import dependence or increase the short-run flexibility of consumption or production, the estimated direct costs of the embargoes are as follows:

	Scenario III: 1980 Embargo	Scenario IV: 1990 Embargo
Price during embargo (1973 dollars)	31.00	31.00
Cost impact (billions of 1973 dollars)	20.7	62.2
Present value of cost impact at 10 percent (billions of 1973 dollars)	14.1	16.4

These figures suggest that the direct costs of production cutbacks are likely to be substantial, especially in the very short run, when demand is inelastic and domestic production cannot increase in response to higher prices. In the longer run, consumption and production may be able to adjust to these higher prices, but the short-run costs of unexpected production cutbacks are substantial. These estimates include only direct costs, excluding the more conjectural but potentially more substantial macroeconomic effects. The political and psychological trauma inflicted by such disruptions cannot be indicated readily by economic-cost measures. However, these very large potential costs indicate that, if policymakers determine that there is a substantial risk of future embargoes, current investments in the billions of dollars may be justified. For example, if the likelihood of a 1980 embargo is one chance in ten, current investments or costs of at least $1.41 billion could be justified if they would spare the economy the direct costs of the embargo impact. An investment of two to five times this amount could be justified if the indirect costs are of similar magnitude to those that may have been incurred during the 1973-74 crisis.

Stockpiling for Future Crises in Oil Supply

Stockpiled oil was assumed to be sold to the domestic market, and exports to other countries were assumed to be prevented by government policy. In actual crises, in line with current agreements, the U.S. government might choose to sell or give oil to more import-dependent allies, such as Japan.

Benefits and Costs of Stockpiles and Optimal Levels of Stocks

The initial benefit of a stockpile release (that is, the net benefit from the first barrel released) lies in the oil price—the value of oil to consumers. The costs per barrel of oil, if it is stockpiled, include original cost plus interest and storage costs. As more is released, after a point, the price will fall and the value to consumers decline.

Efficient-stockpile calculations were carried out under a variety of assumptions concerning the magnitude of storage costs, the indirect costs of an embargo, and the price at which imported oil would be available. Results are presented for both high and low storage and interest costs. The high costs correspond to storage in steel tanks, while the low costs imply that storage costs are zero. The two scenarios therefore bracket the likely merging of storage costs (involving salt domes at the lower end and steel tanks at the high end). Table 4.3 shows gross benefits and net benefits (or cost, if negative) of initial stockpile releases, without accounting for the indirect costs of an embargo.

Stockpiling does not pay in Scenarios I and II since the even very small amounts of stockpiled oil is sold for no more than its total acquisition cost. While the stockpiled oil would be valuable during a 1990 embargo, accumulated carrying costs would greatly exceed the benefits. Net benefits are positive for a 1980 embargo. A stockpile of 2.1 million barrels per day, or 767 million barrels over the year, would yield maximum net benefits. Net benefits total between \$2.3 million and \$20.4 million per day, between roughly \$840 million and \$7.5 billion for the embargo year, depending on storage and interest charges (in 1973 dollars).

TABLE 4.3

Gross Benefits and Net Benefits of Initial Stockpile Release, Assuming No Indirect Costs of an Embargo (1973 dollars per barrel)

Benefit	OPEC Collapses[a]	OPEC Stable[a]	OPEC Stable, 1980 Embargo	OPEC Stable, 1990 Embargo
Gross	4.00	12.73	31.00	31.00
Net				
Low storage costs[b]	-17.28	-8.55	9.72	-26.42
High storage costs[c]	-25.92	-17.19	1.08	-71.20

[a]Net benefits are calculated under the assumption that the stockpile is liquidated in 1980.

[b]Annual storage and interest costs are assumed to equal \$1.273 per barrel, and are compounded from 1975 until the stockpile is liquidated.

[c]Annual storage and interest costs are assumed to equal \$2.56 per barrel, and are compounded from 1975 until the stockpile is liquidated.

Source: Charles River Associates, Policy Implications of Producer Country Supply Restrictions: Overview and Summary (Cambridge, Mass.: CRA, 1976), p. 66.

Taking account of macroeconomic costs and political consequences could justify much larger stocks. The per-barrel indirect costs for the 1973-74 crisis may have ranged between \$23 and \$49 (in 1973 dollars); however, the high estimates seem unrealistic. An indirect cost of \$49 per barrel is more than four times the estimated per-barrel direct cost of the 1973-74 embargo. Even the low estimates of indirect costs are more than twice the direct cost estimates. Efficient stockpiles corresponding to per-barrel indirect costs of \$24.51 are presented in Table 4.4. Including an estimate

TABLE 4.4

Summary of Efficient Stockpile Levels for Future Embargoes

Indirect Costs (per barrel)	Storage/ Interest Costs	1980		1990	
		Million Barrels	Months of 1980 Consumption	Million Barrels	Months of 1990 Consumption
$0	A	767	1.4	0	0
	B	767	1.4	0	0
$24.51	A	1,237	2.3	0	0
	B	1,435	2.7	0	0

Note: Optimal 1990 stockpile is zero in all cases because of accumulated carrying costs.

A: High storage plus interest costs ($2.56 per barrel, per year), compounded to embargo year.

B: Low storage plus interest costs ($1.273 per barrel, per year), compounded to embargo year.

Source: Charles River Associates, Policy Implications of Producer Country Supply Restrictions: Overview and Summary (Cambridge, Mass.: CRA, 1976), p. 67.

of the indirect costs of an embargo increases the size of the efficient stockpile for a 1980 embargo by 60 to 85 percent, from 767 million barrels (2.1 million barrels per day) to over 1.2 billion or 1.4 billion barrels (3.4 or 3.9 million barrels per day), depending on storage and interest costs. Stockpiling becomes a beneficial policy for dealing with a 1990 embargo only if indirect costs per barrel are extremely high.

In general, a stockpile policy could yield positive expected net benefits only if an embargo is considered quite likely. The estimated costs and benefits of stockpiling can be used to determine what probability of an embargo is required for the net benefits of stockpiling to be positive. If indirect costs are negligible for a 1980 embargo, the probability must be at least .69 if storage and interest costs are low; if storage costs are high, the probability must be over .97. That is, if storage costs are at the high end of our range, then an embargo must be almost certain to occur if stockpiling is to be an economically beneficial policy. If storage costs are at the low end of our range, an embargo must be very likely for stockpiling to yield net economic benefits. Allowing for indirect costs of course reduces these "break-even" probabilities significantly.

The efficient stockpiles determined above range from 1.4 to 2.7 months' consumption, at expected 1980 rates; these stockpiles are efficient for an embargo equal in severity to the 1973-74 embargo, lasting a full year. The stockpiles corresponding to an embargo of four months' duration (such as the 1973-74 embargo) would be equivalent to .4-to-.7 months' consumption.

It is possible to justify a U.S. petroleum stockpile equal to two to three months of U.S. consumption without being as pessimistic about the likelihood of a disruption as in the above analysis. As discussed further in the Appendix to this volume, much larger stockpiles can be justified than would otherwise be the case if those larger stockpiles decrease the probability that a disruption will occur in the first place—that is, if they have deterrence effects. This probability-deterrence effect of stockpiles may be significant for disruption scenarios similar to the ostensibly politically motivated embargo of petroleum exports to the United States in 1973. Larger stockpiles would have an even greater deterrence effect against any attempt at short-run price gouging by OPEC.

The generalized policy model described in the Appendix also recognizes price-deterrence effects, whereby U.S. stock releases during disruptions lower the price on the world market and, hence, lower the price of imports for U.S. consumers (and for other consuming nations whose welfare is important to the United States). Price-deterrence effects reinforce probability-deterrence effects; together (under moderate assumptions) they can justify a U.S.

stockpile equal to two to three months of U.S. consumption, even where the probability that a disruption will start in the course of a year is less than 0.06 (equivalent to a probability of 0.46 over the course of a decade).

Policy Merits and Demerits of Stockpiling

Stockpiling is a costly policy option, but it has significant advantages relative to the alternatives we evaluated. It creates fewer substantial vested interests and patterns of dependence than do quotas, tariffs, subsidies, or price supports, and the income transfers from consumers to oil companies are much smaller than for subsidies, tariffs, or quotas. It provides the potential for rapid domestic supply response in the event of an embargo, and consequently is a more potent deterrent to embargoes than are tariffs, quotas, or subsidies, which merely reduce the normal level of imports. This quick-response capability means that consumption can be more fully sustained, imports more effectively reduced, and political and psychological trauma reduced. In principle, stockpiling is reversible and can be implemented without major market adjustments or implementation difficulties.

Although the costs in the event of cartel weakening or collapse are mitigated by the fact that the stockpiled inventory can be sold, the experience of strategic stockpiling policy in the United States indicates that it is difficult to obtain congressional clearance for disposals. Stockpile releases will be surrounded by controversy, and much of the theoretical flexibility may be lost in controversy over who is to get how much of a stockpiled commodity, and at what cost. Private stockholding may be offset, and firms' incentives to avoid risky supply sources may be lessened by expectations that government stocks will bail them out. In addition, stockpiling involves large direct budgetary outlays by the federal government and so may be viewed as politically more costly than other, less beneficial policies.

Tariffs and Quotas to Reduce Crisis Impacts

Under Scenarios I and II, there is no embargo to justify the costs of import restrictions. For a 1980 and a 1990 embargo, we calculated the optimal tariffs and quotas, namely, the levels which produced the most favorable balance between reduction of crisis costs and costs of import restriction in noncrisis years. For a 1980 embargo, we assume that the tariff is imposed in 1975 and lifted after 1980. For a 1990 embargo scenario, we assume that

TABLE 4.5

Basic Elements Used in Calculating the Optimal Tariff for Scenario III (1980 Embargo)

Element	Data		
Embargo price	$31.00 per barrel, 1973 dollars		
Interest rate	10 percent per year in real terms, compounded annually		
U.S. supply curve	Elasticity (6-year average)	=	0.09
	Slope (6-year average)	=	30.7 (million barrels per dollar)
	Slope (1980)	=	87.0 (million barrels per dollar)
U.S. demand curve	Elasticity (6-year average)	=	-0.59
	Slope (6-year average)	=	-290.6 (million barrels per dollar)

Source: Charles River Associates, Policy Implications of Producer Country Supply Restrictions: Overview and Summary (Cambridge, Mass.: CRA, 1976), p. 70.

TABLE 4.6

Basic Elements Used in Calculating the Optimal Tariff for Scenario IV (1990 Embargo)

Element	Data		
Embargo price	$31.00 (1973 dollars)		
Interest rate	10 percent per year, in real terms, compounded annually		
U.S. supply curve	Elasticity (16-year average)	=	0.323
	Slope (16-year average)	=	123.6
	Slope (1990)	=	263.2
U.S. demand curve	Elasticity (16-year average)	=	-0.676
	Slope (16-year average)	=	-383.6

Note: Forecasted production and consumption are obtained by linear interpolation of values in tables 5-15 and 5-16 of vol. 7 of the ETIP study.

Source: Charles River Associates, Policy Implications of Producer Country Supply Restrictions: Overview and Summary (Cambridge, Mass.: CRA, 1976), p. 71.

the tariff (or quota) is in effect from 1975 to 1990. The basic data used in these calculations are shown in Tables 4.5 and 4.6, respectively.

For Scenario I, in which OPEC is assumed to become ineffective by 1980, we analyze two trade policies. In the first version we evaluate the tariff policy found optimal for Scenario IV, a 1990 embargo, while in the second version we evaluate the quota found optimal for Scenario IV. Clearly, neither of these policies is optimal for either Scenario I or II, since with no embargo, the optimal level for a tariff is zero and imports should be unrestricted by a quota. Because actual events are of course highly uncertain, these evaluations are important to weigh the downside risks associated with the options. Tariffs and quotas are considered separately since tariffs allow greater access to lower-priced imports after 1979.

For Scenario II, in which OPEC is stable, we evaluate both the tariff found optimal for Scenario III, assuming that the tariff is removed after 1980, and the optimal tariff for Scenario IV, in which the tariff or quota is assumed to remain in effect throughout the 1975-90 period.

Benefits, Costs, and Redistributive Effects

Table 4.7 shows the costs, transfers, and benefits of tariffs and quotas under the different scenarios. Costs are, of course, greater when the price of crude oil falls, and greatest when access to cheaper oil is restricted by a quota.

If the embargo occurs as expected, the optimal tariff and quota yield positive net benefits. Under Scenario III, the optimal tariff is $3.02 per barrel, or less than 25 percent of the price of crude oil, while under Scenario IV, at $0.88 per barrel, the optimal tariff is less than 10 percent of the crude oil price.

Because of the relative inelasticity of demand and supply, transfer payments to producers and taxpayers are many times larger than the real resource costs of these policies. For example, under Scenario IV, the accumulated transfer payments to producers, including interest, total $149.6 billion by 1990, and transfer payments to the U.S. treasury reach an estimated $54.69 billion. Under a quota, the transfer to the treasury is assumed to result from the sale of import rights. If a quota policy were administered as it has been in the past, rights would be given away to refiners, and transfers to producers would be increased greatly. By contrast, the real resource costs over the 1975-90 period, including interest, amount to only $6.26 billion. Consumers may be particularly unhappy about transfers of such magnitude to the oil companies, an average of $9.35 billion per year, counting interest accrued.

TABLE 4.7

Costs and Benefits of Quotas and Tariffs under the Different Scenarios

Item	Scenario					
	I(A)	I(B)	II(A)	II(B)	III	IV
Tariff[a]	0.88	—	3.02	0.88	3.02	0.88
Quota[b]	—	2,164	6,011	2,164	6,011	2,164
Transfer to producers[c]	118,006	163,865	100,434	149,606	100,434	149,606
Transfer to treasury[c]	85,532	94,053	23,549	54,687	23,549	54,687
Resource cost—total[c]	22,958	42,849	11,468	7,054	9,831	6,858
Resource cost per year[c]	1,412	2,678	1,911	441	1,639	457
Gross benefits—total[c]	—	—	—	—	23,532	21,971
Net benefits[c]	-22,958	-42,849	-11,468	-7,054	13,701	15,113

[a]Dollars per barrel.

[b]Million barrels per year.

[c]Millions of 1973 dollars.

Scenario descriptions

I: OPEC collapses by 1980; real price remains at $4 per barrel, 1980-90; (A) optimal tariff for a 1990 embargo remains in effect from 1975 to 1990; (B) optimal quota for a 1990 embargo remains in effect from 1975 to 1990.

II: OPEC is stable; real price remains at $12.73 per barrel from 1975 to 1990; (A) optimal tariff (quota) for a 1980 embargo is in effect until 1980 and is then removed; (B) optimal tariff (quota) for a 1990 embargo is in effect from 1975 to 1990.

III: OPEC is stable; embargo occurs in 1980; optimal tariff (quota) in effect from 1975 to 1980 and removed after the embargo.

IV: OPEC is stable; embargo occurs in 1990; optimal tariff (quota) in effect from 1975 to 1990.

Source: Charles River Associates, Policy Implications of Producer Country Supply Restrictions: Overview and Summary (Cambridge, Mass.: CRA, 1976), pp. 72-73.

Policy Merits and Demerits of Tariffs and Quotas

Tariffs and quotas make no sense unless a heavy weight is placed on the embargo contingency. The costs incurred in nonembargo scenarios are large, on the order of magnitude of those for stockpiling, given the roughness of the estimates. In addition, these costs are compounded by redistribution and the creation of vested interests.

Tariffs and quotas have some additional advantages and drawbacks. Rather than requiring direct outlays from the federal budget, as stockpiling does, tariffs are a potential source of revenue for the government that could be used to ease the distributional impact of increased oil prices by providing tax relief. Quotas also do not require direct budgetary outlays, but unlike tariffs, they generate no government revenue unless the quota tickets are sold. In fact, experience indicates import rights might be given away.

Both tariffs and quotas pose potentially serious problems of reversibility, because they encourage the expansion of protected domestic petroleum production dependent on the tariff or quota. To be put into effect, both policies also imply substantial market adjustments. Users will experience increased energy costs; there is some equity in this, as the costs of financing the reduction in import dependence should be borne by energy users.

While tariffs are preferable if prices fall, quota rights could be auctioned off, perhaps leading to increased competition within OPEC and a more likely collapse of cartel discipline. A quota provides domestic producers with greater certainty of an assured market, but it is not as flexible a device as a tariff with respect to price decreases by OPEC. Whereas domestic consumers will benefit from lower prices in the event of a tariff, a quota restricts access to the imported supply, regardless of its price.

Subsidies for Domestic Oil Production

Since, in the absence of an embargo or a supply cutoff, the optimal subsidy is zero, for Scenario I, in which OPEC is unstable, we consider a subsidy optimal for a 1990 embargo. For Scenario II, in which OPEC is stable, we consider the subsidies optimal for both a 1980 and a 1990 embargo. As with the analysis of tariffs and quotas, these calculations aid in assessing policy risks.

Benefits, Costs, and Income Redistribution

The results of a subsidy policy are summarized in Table 4.8. Several interesting features of this table stand out. First, if the

TABLE 4.8

Costs and Benefits of Domestic Subsidies under the Different Scenarios

Item	Scenario				
	I	II(A)	II(B)	III	IV
Subsidy[a]	1.06	6.89	1.06	6.89	1.06
Resource costs[b]	4,522	5,620	2,494	5,620	2,494
Average costs per year[b]	283	937	156	937	156
Transfers to producers[b]	142,912	232,294	180,689	232,294	180,689
Average transfer per year[b]	8,932	38,716	11,293	38,716	11,293
Benefits[b]	—	—	—	11,245	4,966
Net benefits[b]	-4,522	-5,620	-2,494	5,625	2,472
Average net benefits per year[b]	-283	-938	-156	938	155
Additional average output per year[c]	237.5	212	131	212	131

[a]Dollars per barrel (1973).

[b]Millions of 1973 dollars.

[c]Million barrels.

Scenario descriptions

I. OPEC collapses by 1980; real price remains at $4 per barrel, 1980-90; optimal subsidy for a 1990 embargo remains in effect from 1975 to 1990.

II. OPEC is stable; real price remains at $12.73 per barrel from 1975 to 1990; (A) optimal subsidy for a 1980 embargo is in effect until 1980 and is then removed; (B) optimal subsidy for a 1990 embargo is in effect from 1975 to 1990.

III. OPEC is stable; embargo occurs in 1980; optimal subsidy in effect from 1975 to 1980 and removed after the embargo.

IV. OPEC is stable; embargo occurs in 1990; optimal subsidy in effect from 1975 to 1990.

Source: Charles River Associates, Policy Implications of Producer Country Supply Restrictions: Overview and Summary (Cambridge, Mass.: CRA, 1976), pp. 76-77.

embargo is expected to occur in 1980 rather than in 1990, the optimal subsidy is almost six times as great—over 50 percent of the oil price. This difference arises because the longer the subsidy must be in effect, the higher the costs are, so that the optimal subsidy that would be in effect 15 years before actual benefits are achieved during an embargo is lower than the subsidy that would be in effect only five years.

Second, because of a higher subsidy per barrel, yearly transfers to domestic crude oil producers are much greater under Scenario III than Scenario IV. That is, total transfers over the six-year period are larger than the transfers over the 16-year period, taking into account, in both cases, accumulated interest.

Third, subsidies yield lower benefits than either quotas or tariffs, assuming that a supply cutoff occurs, because the quotas or tariffs affect consumption as well as production. Under Scenario III, for example, the net benefits of an optimal quota or tariff are $13.7 billion, while the net benefits of an optimal subsidy are only $5.6 billion. Correspondingly, however, the net costs of a quota or tariff are much higher than those of a subsidy if the expected supply cutoff or embargo fails to occur.

Fourth, the transfer payments of domestic subsidies are extremely large relative to the costs and benefits. It is quite probable that taxpayers would resent making such large payments to oil producers.

Policy Merits and Demerits of Subsidies for Domestic Crude Oil Production

Like stockpiling, tariffs, and quotas, subsidies are cost effective only if very heavy weight is given to the embargo contingencies. Unlike tariffs and quotas, subsidies do not encourage energy conservation, and like stockpiling, subsidies involve substantial federal outlays. Because subsidies to domestic producers would create potential domestic capacity for increased quick supply response during an embargo, the trauma of an embargo is likely to be somewhat less under a policy of subsidies than under some other policy alternatives. The redistributive effects and government expenditures are large; energy consumers and producers would gain at the expense of taxpayers. The total subsidy bill could be reduced by supporting only selected higher-cost producers, such as those using advanced recovery techniques or those from new producing regions, though this could result in foregoing output, or by subsidizing production increases over some base.

The subsidy analysis assumes that domestic producers will respond to the higher effective prices as they have in the past, and that the risk of substantial price declines due to an OPEC collapse will not deter investment. If this risk does deter investment, greater subsidies and larger transfers to producers would be required to yield the benefits shown. Subsidies implemented as price guarantees might mitigate this problem, though at the cost of continuing to encourage high domestic production if import prices do fall, as in Scenario I.

The cost of subsidy programs is visible, unlike that of tariffs and quotas. This is desirable on accountability grounds. The required increases in the federal budget may be difficult to obtain.

Synthetic-Fuels Development Options

Large-scale development of a synthetic-fuels industry in the United States before 1990 is not likely without changes in federal policy. As of 1975 only three companies had filed applications with the Federal Power Commission indicating firm plans to open plants for the production of high-Btu gas from coal. While a number of groups appeared capable as of the mid-1970s of developing commercial-scale oil shale-processing facilities by 1985, the only active project, the Colony Development Operation, has postponed construction indefinitely. There are no significant low-Btu gas facilities and no coal-liquefaction plants in the United States. Of the synthetic fuels under consideration, only synthetic crude oil from Canadian tar sands has reached the stage of commercial production, but Canadian energy policy makes this option infeasible.

The slow pace of synthetic-fuels development is due to a number of factors. First, commercialization of many processes entails great risk. Currently available coal-conversion processes may be obsolete by the time new plants could go on-stream, while new technologies are not proven yet. Technical uncertainties haunt all new synthetic-fuels processes, while proven coal-conversion technologies are of limited usefulness with U.S. coal.

Second, even in the absence of technical difficulties, uncertainty about prices of synthetic products, conventional fuels, and process feedstock would deter development. Since future oil prices are unknown, investors do not know the price at which their product could be sold. Similarly, uncertainty about future regulatory policy makes it difficult to predict both the price at which substitutes for natural gas can be sold, and the possible regulatory constraints that could be imposed upon producers of synthetic natural gas. Finally, coal prices have increased greatly in the last two years; coal costs account for a quarter to a third of product cost for some conversion technologies.

Mining of subbituminous coal, oil shale, and tar sands involves potential environmental-degradation problems as well as rapid development of sparsely settled areas of the western United States and Canada. As a result, opposition from environmental groups and state governments threatens to flow or halt synthetic-fuels development.

Presumably as a result of these and other problems, few private investors have been willing to undertake the risk of building synthetic-fuels plants. Much of the research and development of new processes has been supported by the federal government. Commercialization of natural gas substitutes would require heavy investment by a gas industry that might well find Alaskan natural gas via pipeline a more attractive source of new supply between now and 1981. Without government support, many synthetics projects are not likely to be developed.

Policy Alternatives

Development of a synthetic-fuels industry will require changes in government policy. The cost-benefit analysis presented later indicates that, even without allowing for the high risk of these ventures, and considering only private costs, under 1976 economic conditions and known technology several processes would not even yield a 10 percent real rate of return. All processes would sustain losses if oil prices, and hence product prices, fell, even moderately. In light of this analysis and the constraints on development discussed in previous sections, alternative government policies for promoting development should be considered.

No Government Action. The first option for the government is to do nothing. This action would be reasonable if it is judged that private investors correctly evaluate the risks involved, if there are no overriding noneconomic goals for synthetic-fuels development, and if there are no institutional obstacles that may thwart efficient development.

Capital Subsidy. A second government option would be to subsidize the capital expenditures of a synthetic-fuels industry. Capital costs are the major component of project costs. These costs increased 26 percent from 1973 to mid-1975 alone.

If a subsidy is to be given, a capital subsidy to synthetic-fuels producers has certain attractive features. The federal commitment is clearly limited and not dependent on the level of world oil prices. The amount of assistance can be set at any level desired, and since it can be set at a fixed amount or a percentage of plant cost, the federal outlay is predictable.

Most capital-subsidy schemes are administratively simple. Other alternatives, in order of increasing cost to the federal government for any given level of support, include federal loan guarantees, direct federal loans, and capital grants to producers.

Price Supports. Under a third option, the government would deal directly with the issue of uncertainty about future product prices by guaranteeing that a synthetic-fuels industry will obtain some minimum price for its output. This policy does not provide a general floor for all oil prices but only for synthetic fuels. The policy could be designed to apply to as broad or as narrow a group of firms and products as would be desired. Some alternatives even provide an opportunity for the government to make a profit.

Problems with a price-support program include its unpredictable cost to the government. Federal outlays can be forecast only if future oil prices and future product costs can be estimated. Further, with this approach, federal outlays depend on the world price of oil. Guaranteeing a market for output at some reasonable price may reduce market pressure on firms to produce efficiently. A continuing subsidy of operating costs such as that provided by a price support might create an infant industry dependent on government assistance, and might create a vested interest in continuation of the support. However, the price support need not be pegged at levels that produce a high level of expected subsidy. A support level of $9 or $10 per barrel would probably eliminate the main deterrent effect of the risk that petroleum prices will collapse or be manipulated by OPEC to damage the synthetics industry. The subsidy could be designed for automatic phaseout over a specified time period. This approach would produce net disbenefits only in the event of a relatively early OPEC collapse combined with active introduction of the synthetic fuels eligible for price support.

Government-Owned Plant. Production of synthetic fuels in a government-owned plant entails the largest direct commitment of federal resources for any given output level of all the policy options. It solves the capital-availability problem but gives no incentives to private ownership or large-scale commercial development, except as a pilot project. This scheme escapes criticism as a giveaway to energy companies.

Cost-Benefit Analysis of Synthetic-Fuels Options

Because of uncertainty about potential development scale by 1990, we evaluated all of the options on an annualized per-barrel basis. The per-barrel benefits are simply the value of oil displaced—the price of crude under the alternative scenarios.

For purposes of initial cost measurement, we have taken estimates of the Synfuels Interagency Task Force as the best available benchmark figures, even though government estimates of the prospective costs of new technological processes often tend to be overly optimistic. Thus, our analysis is based on information available in 1975 on synthetic fuels options. Because of the rapidly changing technologies and economic conditions our analysis should be interpreted as illustrative only for evaluating policies as of 1980 and later years.

We considered direct costs that must be borne by private operators; environmental costs falling partly on producers and partly on society; and, as infrastructure costs, the added public expenditures needed for the rapid industrial and population growth in sparsely settled areas that is critical to production of most synthetics. Direct cost estimates were based on capacity operation of feasible plant mixes. We assumed a required after-tax rate of return of 10 percent in real terms on investments in synthetics.

Table 4.9 shows the per-barrel costs of the options considered. Synthetic natural gas and syncrude are by far the most expensive of the different processes considered. The bottom-line entries in Table 4.9 can be interpreted as the minimum product price required by the plants of each program for them to earn a 10 percent real rate of return on investment. By this criterion, synthetic natural gas and syncrude from coal are the least attractive alternatives, while oil from tar sands is the most attractive. Tar-sands costs do not include any export levies or environmental costs, and, as the economic deposits are in Canada, tar sands are not an attractive alternative to reduce U.S. dependence on imports.

Among domestic processes, low-Btu gas and oil shale are the most attractive. In fact, the low cost estimate of these two processes raises the question of why private enterprise is not moving more aggressively to implement these technologies. Since the corporate sector does not lack the information, interest, or sophistication to perceive major commercial opportunities in the energy field, the reasons apparently lie in a combination of the following factors. First, published cost estimates of new technologies are highly uncertain and tend to be optimistic. The history of cost estimates for synthetic fuels has been one of repeated upward revision. Second, commercial development of shale oil faces serious impediments in the acquisition of water rights, the use of federal lands, and opposition on environmental grounds. Also, commercial developers may be concerned with the possibility that OPEC will weaken and that the oil price will fall. Given the obstacles and uncertainties involved,

TABLE 4.9

Synthetic-Fuels Costs: Minimum Acceptable Prices to Guarantee 10 Percent Rate of Return, 1976-90 (using cost estimates available in 1975)

Cost	Oil Shale (range)	Oil Shale (weighted average)	Tar Sands	Synthetic Natural Gas	Low-Btu Gas	Syncrude (range)	Syncrude (weighted average)
Direct	$ 7.34-10.23	$7.82	$4.72	$13.58 (2.34)	$6.29	$14.29-16.30	$15.77
Environmental	0.44	0.44	NA	0.44 (0.08)	0.52	0.40-0.42	0.42
Infrastructure	0.29	0.29	1.24	2.08 (0.36)	1.53	0.55	0.28
Total cost (minimum acceptable price)	8.07-10.96	8.55	5.96	16.10 (2.78)	8.34	15.24-17.27	16.47

Notes: All main entries are in 1973 dollars per barrel of oil equivalent. Entries in parentheses are in 1973 dollars per million Btu's (or per thousand cubic feet) and assume 5.8 million Btu's per barrel, oil equivalent.

Price ranges presented for oil shale and syncrude are based on low and high cost estimates for different technologies. Low-cost technologies are surface retorting (oil shale) and H-Coal (syncrude); high-cost technologies are in situ retorting at current cost estimates (oil shale) and Fischer-Tropsch (syncrude). Weighted averages are based on 1990 shares of output from table 6-8 of vol. 7 of the ETIP study done by Charles River Associates.

Capital-cost components of direct and environmental costs assume a 10 percent interest rate and a 25-year plant life; the implied capital-recovery factor is 11.0168 percent. Operating-cost components of direct and environmental costs are calculated for plants of capacities given in table 6-8 of vol. 7 of the ETIP study, assuming 330 stream days per year.

Environmental and infrastructure costs are based on Synfuels Interagency Task Force estimates.

Special assumptions (all figures are converted to 1973 dollars):

1. Oil shale—mining costs are included for 30-gallon-per-ton shale; investment in these plants of $443.26 million per plant; annual operating costs of $72.30 million per plant. One plant uses in situ retorting of 18-gallon-per-ton shale; investment in this plant of $373.74 million; annual operating costs of $127.9 million.
2. Tar sands—investment of $589.57 million per plant; annual income taxes and royalties of $50.9 million per plant; other operating costs, $79.02 million per plant.
3. Synthetic natural gas—western coal is used at $5.43 per ton. Coal-mining and other support facilities are not included in direct plant cost; investment of $580.39 million per plant; annual operating costs of $128.86 million per plant.
4. Low-Btu gas—eastern coal is used at $10.08 per ton; investment of $105.19 million per plant; annual operating cost of $58.23 million per plant.
5. Syncrude—Fischer-Tropsch plant uses western coal at $5.43 per ton; investment of $1.17 billion per plant; annual operating cost of $206.64 million per plant. H-Coal plant uses eastern coal at $10.08 per ton; investment of $271.65 million per plant; annual operating cost of $77.28 million per plant.

Sources: Synfuels Interagency Task Force, Recommendations, vols. 2-3 (Washington, D.C.: Government Printing Office, June 1975); Organization of Economic Cooperation and Development, Energy Prospects to 1985, vol. 2 (Paris: OECD, The Petroleum Economist, March 1975).

the risks are high. In addition, there may be a tendency to defer development initiatives in the hope that a subsidy will be legislated to make the initiative more profitable. Thus, the uncertain anticipation of subsidy may actually deter development. When all of these factors are taken into account, it is not surprising that there is no large-scale effort underway to develop shale oil and low-Btu gas.

The comparatively low cost estimates for low-Btu gas are particularly puzzling in view of the current behavior of gas-pipelines operators and gas distributors who are investing in seemingly higher-cost Liquefied Natural Gas (LNG) and Lurgi coal-gasification facilities. The most plausible reasons for such interest may be that the cost estimates we have used are too low, and that the Lurgi gas and LNG processes offer a more concrete source of supply for relieving gas shortages by virtue of using received technologies that have a better experience base than that of low-Btu gas.

The cost estimates do not take account of any possible cost reductions that might result from the industry's added experience after each technology reaches commercial scale in 1985; nor is there any allowance for the cost-increasing effect of higher levels of development, increased pressure on water resources, higher levels of environmental-protection effort, or greater infrastructure costs. In effect, we assume a one-for-one offset of these cost-reducing and cost-increasing effects, although these cost uncertainties can greatly affect the economic and policy merits of the alternatives. For example, if operators are liable for all reported environmental and infrastructure costs, economic viability of synthetic natural gas would require a cost decrease of 42 percent; if operators are liable only for direct costs, the required decrease is 25 percent. For syncrude the equivalent required cost decreases are 44 percent and 38 percent, respectively. Experience in commercial-operations equipment design and construction might produce these reductions, as could a combination of decreased coal prices and capital costs due to lower interest rates; however, heretofore unknown difficulties could escalate costs. On the other hand, the declines in oil prices that would eliminate the commercial viability of oil shale and low-Btu gas are 8 percent and 12 percent, respectively, if operators must cover total costs. The equivalent percentages are 19 and 43 percent if operators must cover only direct costs. Equivalent cost escalation could destroy commercial viability. It thus appears that if operators must pay for all costs (that is, if environmental requirements are not relaxed), oil shale and low-Btu gas are quite vulnerable both to cost escalation and risk of lower oil prices.

Net Benefits of Synthetic-Fuel Options

Table 4.10 presents the per-barrel annual subsidies required for each synthetic-fuel technology to break even in each scenario. A zero subsidy indicates that an option yields positive net benefits under the particular scenario; that is, the value of imported oil displaced at least covers total costs (including possible losses in nonembargo years). Clearly, if the cost numbers are correct, oil shale and low-Btu gas should be encouraged if OPEC is expected to be stable. On the other hand, if OPEC should collapse, both oil shale and low-Btu gas would require large annual subsidies.

TABLE 4.10

Annual Per-Barrel Subsidy Payments Required for Each Synthetic-Fuel Technology to Break Even in Each Scenario (1973 dollars per barrel, oil equivalent)

Scenario	Oil Shale	Synthetic Natural Gas	Low-Btu Gas	Syncrude
I	4.55	12.10	4.34	12.74
II	0	3.37	0	4.01
III	0	3.37	0	4.01
IV	0	2.77	0	3.38

Note: Subsidy is per barrel of oil equivalent to guarantee 10 percent rate of return. In Scenario IV it is assumed that the private firms capture the full social benefits of extra production capacity during the 1990 embargo.

Source: Compiled by the authors from data shown in Table 4.9.

An interesting question arises as to whether the benefits of increased domestic production capacity during a potential future embargo outweigh the losses in nonembargo years for the high-cost synthetic fuels. It should be noted that none of the alternatives considered is expected to yield significant output before 1982, and consequently none would provide benefits during a 1980 embargo.

Under the assumptions of our analysis, even if producers can capture the full direct economic value of the extra production capacity during a 1990 embargo (not including deterrence effects and indirect benefits), they would still need an annual subsidy of $2.74 per barrel in the case of synthetic natural gas, and $3.38 in the case of syncrude, to break even, assuming a 10 percent real after-tax rate of return over the 1976-90 period (in 1973 dollars). In other words, the contingency benefits of the extra capacity during an embargo are substantially outweighed by the extra production costs in nonembargo years.

An approximation of the per-barrel capital-outlay subsidy required to induce private development for each scenario can be obtained by multiplying the per-barrel subsidies by a factor of 9.1. One million barrels of oil shale capacity requires a subsidy of about $41.4 million if OPEC collapses in 1980.

It would be extremely expensive to induce development of synthetic natural gas or syncrude on a substantial scale by 1990. We conclude that synthetic natural gas and syncrude ought not to be subsidized heavily for the purpose of reducing the impact of oil-supply disruptions.

The benchmark costs of oil shale and low-Btu gas are below the assumed price of oil in the scenario of OPEC stability. If these costs can be realized, our assumptions indicate that production capacity for these synthetics ought to be expanded rapidly after 1985. Such expansion would probably apply upward pressure on costs and would encounter bottlenecks in equipment supply, water, and resistance on environmental grounds; nevertheless, it is clear that if these cost estimates are credible, the economically efficient level of development will be higher than we have assumed in this analysis for purposes of calculating per-barrel benefits and costs.

Policy Considerations

A policy of encouraging accelerated development of the more promising synthetic fuels—shale oil and low-Btu gas—encounters some significant difficulties. The costs are highly uncertain and probably will be significantly above our benchmark estimates. A moderate decline in expected oil prices or a moderately higher cost of development would eliminate the net benefits of subsidies for synthetics and delay or abort their actual development; there also is a substantial risk that costs and implementation difficulties will be much greater than forecast. Subsidy programs requiring continued expenditure to assure commercial viability could consequently involve a highly uncertain, open-ended, and potentially very large commitment of federal funds.

Such a commitment carries with it the problems of creating a vested interest. Depending on the structure of a subsidy program, it may prove difficult to wean the infant industry from federal aid or preserve healthy commercial incentives to minimize costs. If synthetic-fuels technologies are capable of achieving economic viability over the long term, then they should not require subsidization once the risks and obstacles that accompany their early development have been removed.

Importance of Risk Reduction and Policy Coordination

The above considerations and the nature of obstacles to future development suggest both that federal policy be concentrated on approaches that reduce the risks and obstacles with relatively low commitment of federal funds, and that programs that invite an open-ended chain of commitments of subsidy funds be avoided. From this standpoint, a temporary program of loan guarantees, to expire after 1985, would be preferable to capital grants; price supports would be the least desirable policy.

Issues of water-resource allocation and environmental protection are among the greatest sources of risk and uncertainty deterring development of shale oil and synthetic gas production. Capital grants and price supports do little to reduce these obstacles or reduce the cost of synthetics. Mitigating these obstacles requires a concerted effort by policy makers to achieve a consensus on water and environmental questions and to provide an efficient legal and regulatory environment; such a consensus is needed for investors in these technologies to make reasonable predictions of what they can or must do. The success of such an effort depends more on the political and public administrative proficiency of the government leadership than on heavy public expenditures. Focusing federal research and development efforts on environmental abatement and on technologies that conserve water would also help reduce the environmental and water-resource impediments to development.

Subsidization of synthetics almost certainly involves a redistribution of income to the benefit of producers, but the effect is much less than in the case of subsidies for domestic oil production, or of import quotas, or tariffs. If low-cost, reversible approaches are followed rather than capital grants or high-level price supports, the income-distribution effects would be modest.

An advantage of synthetics development is that it could provide improved long-run alternatives to imported oil without accelerating depletion of dwindling domestic petroleum resources, as U.S. shale and coal resources are enormous. If policy makers believe that OPEC may be stable for some time, the development of such long-term and potentially flexible alternatives becomes highly desirable.

Conversion of Generating Facilities to Coal

Conversion of generating facilities to coal can lower oil imports directly by reducing oil use and indirectly by releasing natural gas from use as utility boiler fuel. This gas can then displace oil in other uses, such as home heating. Both effects can reduce embargo impacts. A major policy instrument of the federal government's effort toward energy independence is the Energy Supply and Environmental Coordination Act of 1974. The act empowered the Federal Energy Administration to require conversion to coal of some existing oil-burning facilities, and to require coal-burning capability for facilities on order. The title of the act correctly portrays the basic issues: environmental-policy constraints and control problems induced utilities, particularly in the Northeast, to shift to oil. Reversing this shift required a counterforce to overcome the conflict and uncertainties surrounding environmental policy.

In performing our analysis of coal conversion, we appraised the feasibility of the FEA's initial program of conversion orders, paying particular attention to its consistency with environmental policy at the federal, state, and local levels. Using realistic assumptions about plant life and capacity factors, we estimated the amount of electricity involved, the coal requirements, and the potential oil displacements. We devoted major attention to identifying possible constraints on coal supply and on the achievement of effective environmental controls. These analyses resulted in estimates of oil displacement and associated costs. The oil displacements then could be valued under the various scenarios, and costs could be subtracted to obtain net benefits. Our principal finding is that coal conversion is economically viable under all the scenarios and does not impose major costs if OPEC collapses (Scenario I).

We found that a feasible base case could generate a cumulative present value in 1976 of net benefits (evaluated in 1973 dollars) ranging from $1.7 billion for Scenario I (OPEC collapse) to about $15.3 billion for Scenario III (a 1980 embargo). A much more optimistic, extended case yielded net benefits ranging from about $3.4 billion (Scenario I) to $26.3 billion (Scenario III). While we believe the programs are feasible, achieving the assumed results will require greater policy coordination and a higher degree of resolution of environmental conflicts than has been achieved to date.

Cases Analyzed

Our benefit-cost analysis of coal conversion is based on two coal-conversion scenarios. The base case, which is realistic to moderately optimistic, assumes that the 1975 FEA prohibition-order program is implemented three years later than FEA's over-ambitious target of substantial implementation by mid-1977. The extended case is an upper bound. It assumes a concerted effort by the federal government to convert all of the 71 plants on the original FEA conversion list, preserving the three-year lag as in the base case.

Evaluation Methodology and Assumptions

Tables 4.11 and 4.12 present the results of the base- and extended-case programs in terms of electricity generation, coal consumption, and oil and gas savings.[2] The detailed assumptions underlying the estimates are presented in the notes to these tables. It should be noted that oil savings, as presented in these tables, do not represent total energy savings due to the programs; the programs themselves use energy. Natural gas savings were converted to barrels of oil equivalent and added to oil savings to obtain total imported-oil savings.

The base case involves conversion of 10,304 megawatts of generating capacity, occurring over the period from 1976 to 1980; and the extended case requires conversion of 20,863 megawatts of capacity over the same period. The base case involves essentially those plants which have received conversion orders under the FEA program, while the extended case includes the total of any potential prohibition-order candidates as announced by the FEA in May 1975.

Conversion involves both significant capital investment and increases in operating costs for the newly converted plants. For the above cases, we have accepted FEA's estimates of the costs involved, $7.50 per kilowatt in additional investment and 0.2 mills per kilowatt-hour (kWh) in additional operating costs after conversion.

The analysis assumes that western coal is available to meet part of the steam coal demand, and a weighted-average high-sulfur coal price was used for all plants, regardless of location. A single low-sulfur coal price was used on the assumption that western low-sulfur coal would be available at a delivered price equal to the forecast price of eastern low-sulfur coal.

The base case assumes that 10 percent of coal consumed by converting plants must be low-sulfur, while the extended case assumes that 40 percent of such coal must be low-sulfur. These estimates are based on the FEA's analysis of the coal-burning capabilities of plants in the two groups and on the options available for meeting pollution-control standards.

TABLE 4.11

Electricity Generation, Oil Imports Displaced, and Coal Requirements: Base-Case Conversions

Year	Electricity Generated (millions of kWh)	Oil Imports Displaced (millions of barrels)	Coal Requirements (millions of tons)
1976	9,930	17.93	4.27
1977	19,562	35.31	8.42
1978	28,896	52.17	12.43
1979	37,932	68.49	16.32
1980	46,768	84.44	20.12
1981	45,381	81.94	19.53
1982	43,988	79.42	18.93
1983	42,697	77.09	18.37
1984	41,406	74.75	17.82
1985	40,115	72.42	17.26
1986	38,925	70.28	16.75
1987	37,735	68.12	16.24
1988	36,641	66.15	15.77
1989	35,547	64.18	15.30
1990	34,458	62.22	14.83

Notes: All entries are totals per year. The assumptions and parameter values used to derive these series are described in the text. Oil imports displaced are barrels of oil equivalent, reflecting both oil and gas savings according to their respective heat content.

Electricity-generation estimates assume conversion of 10,304 MWe at a rate of 2,061 MWe per year, for 1976-80. Electricity is generated with a 48-percent capacity factor when burning oil or gas, and a 55-percent capacity factor in the first year after conversion to coal. Utilization of converted plants decays at a rate of 3 percent per year. Plants operate 8,760 hours per year.

Oil imports displaced reflect both oil and gas displaced by conversion to coal. It is assumed that 69.9 percent of converted generation is from oil-fired plants operated with a heat rate of 10,826 Btu's per kWh, while 30.4 percent of converted generation comes from gas-fired plants operated with a heat rate of 10,733 Btu's per kWh. Total imports displaced assume that 90 percent of gas displaced is used elsewhere to substitute for oil and 1.0 million Btu's per mcf for gas are assumed.

Coal requirements are calculated assuming that plants burn coal with heat content of 23.65 million Btu's per ton, with a heat rate of 10,176 Btu's per kWh.

The above notes apply as well to data in Table 4.12.

Source: Charles River Associates, Policy Implications of Producer Country Supply Restrictions: The World Energy Market (Cambridge, Mass.: CRA, 1976), p. 184.

TABLE 4.12

Electricity Generation, Oil Imports Displaced, and Coal Requirements: Extended-Case Conversions

Year	Electricity Generated (millions of kWh)	Equivalent Oil Imports Displaced (millions of barrels)	Coal Requirements (millions of tons)
1976	20,110	36.31	8.65
1977	39,810	71.51	17.04
1978	58,510	105.63	25.18
1979	76,800	138.65	33.05
1980	94,700	170.98	40.75
1981	91,880	165.88	39.53
1982	89,070	160.81	38.32
1983	86,450	156.08	37.20
1984	83,840	151.37	36.07
1985	81,230	146.66	34.95
1986	78,810	142.28	33.91
1987	76,400	137.93	32.87
1988	74,190	133.94	31.92
1989	71,980	129.95	30.97
1990	69,770	125.96	30.02

Source: Charles River Associates, Policy Implications of Producer Country Supply Restrictions: The World Energy Market (Cambridge, Mass.: CRA, 1976), p. 186.

The Treatment of Environmental-Control Costs

The extended-case plants not included in the base-case list differ from the base-case plants principally in the environmental difficulties involved in their conversion. We have therefore assumed the capital and operating costs for environmental-control equipment to be 25 percent higher for the extended case than for the base case.

Three types of environmental costs were considered. For the capital cost of scrubbers, we have used the FEA's estimate of $80 per kilowatt for the base case. We assume total operating costs for scrubbers and precipitators of four mills per kWh. This figure is higher than the FEA's estimate in its own evaluation of the conversion program, but appears more reasonable than FEA's in the light of actual experience with scrubbers to date. It is roughly the upper bound of an EPA range of estimates of scrubber costs, and roughly the mean of the Commerce Technical Advisory Board's estimate of total scrubber and precipitator costs. Finally, the FEA identified certain plants that needed new or upgraded electrostatic precipitators. We accept the FEA's estimate of $20 per kilowatt for the base case and apply it to all plants that the FEA considered in need of precipitator improvement. The extended case is analyzed in similar fashion, assuming $100 per kilowatt in scrubber capital cost, five mills per kWh in scrubber and precipitator operating cost, and $25 per kilowatt in precipitator capital cost.

The gross benefit of the conversion program is the total value of the oil saved by conversion to coal. This benefit is easily calculated for the four oil-price scenarios. For purposes of this analysis, it was assumed that 90 percent of the natural gas liberated from utility use would displace oil, with the remaining 10 percent displacing coal and electricity for industrial-process heat generation and for space heating. The portion of gas that was assumed to displace oil was then converted from cubic feet to barrels of oil equivalent, and the result was added to oil savings to obtain an estimate of total imported-oil savings. This total, evaluated at assumed oil prices, represents the benefit of the program.

Benefits and Costs

The costs and benefits for the conversion programs were calculated for each year in the 1976-90 period. Capital costs were amortized over a 15-year period, and the resulting cost and benefit series were then capitalized at a 10-percent real discount rate in the same manner as was done for the other policy options.

Table 4.13 presents the benefits and costs of coal conversion. The policy is highly cost effective, both as a means of reducing the present cost of energy and as a means of mitigating the adverse effect of future embargoes. Of all the policy options analyzed, coal conversion stands out as producing substantial positive net benefits for each energy-market scenario.

In calculating the cost of fuel for electric generation in the event of an OPEC collapse in 1980, we assumed that the converted plants would revert to burning residual oil within six months after the collapse. Because of the flexibility of generating equipment with respect to choice of fuel, the conversion of generating facilities to coal does not involve a high net cost in the event of an OPEC collapse.

TABLE 4.13

Costs and Benefits of Coal-Conversion Cases under Various Oil-Price Scenarios (current value in 1976, evaluated in millions of 1973 dollars)

Case	Scenario I	II	III	IV
Base				
Benefits	7,426.0	24,659.4	28,767.9	25,765.6
Costs	5,750.9	13,458.7	13,458.7	13,458.7
Net benefits	1,675.1	11,200.7	15,309.2	12,306.9
Extended case				
Benefits	15,036.2	49,929.5	58,248.7	52,169.6
Costs	11,660.2	31,982.1	31,982.1	31,982.1
Net benefits	3,376.0	17,947.4	26,266.5	20,187.5

Notes: All entries are in millions of 1973 dollars, with interest cumulated at 10 percent to 1990. Rationales for underlying assumptions with respect to coal conversion, coal-market parameters, electric generation, oil displaced, and costs are developed in appendix 4A of vol. 7 of the CRA report for ETIP. Calculation method is detailed in appendix 6A of vol. 7. Parameter values may be summarized as follows: Base case assumes conversion capital cost of $7.50 per kilowatt (KW), environmental capital cost (except electrostatic precipitator [ESP]) of $80 per KW, ESP capital cost (where applicable) of $20 per KW. Extended case assumes conversion capital cost of $7.50 per KW, environmental capital cost (except ESP) of $100 per KW, and ESP capital cost (where applicable) of $25 per KW.

Capital costs for the sinking-fund case are calculated assuming annual interest cost of 10 percent and sinking-fund factor of 3.15 percent for total annual capital charge of 13.15 percent of initial outlay. Sinking-fund factor assumes 10-percent interest rate, 15-year plant life.

All capital costs are discounted at 10 percent to 1976 values.

Sources: Charles Gushee, Financial Capitalization Rate Tables (Boston: 1974), Table 8, p. 726; Charles River Associates, Policy Implications of Producer Country Supply Restrictions: The World Energy Market (Cambridge, Mass.: CRA, 1976), p. 345.

Policy Considerations

Unlike the other policies we have considered, coal conversion does not require an embargo for a clear and substantial positive net benefit to be realized, nor would it precipitate a large net cost in the event of an OPEC collapse. The shift to less costly fuel that the ESECA program of prohibition orders will reflect would occur naturally from the private initiative of electric utilities, were it not for the conflicts, obstacles to coal use, and uncertainties that surround the implementation of environmental policy, especially at the state and local levels. If the environmental ground rules can be resolved so that utilities are assured permission to burn coal, and if the reliability of pollution-control equipment, especially scrubbers for flue-gas desulfurization, can be established, conversion is a worthwhile policy whether an embargo occurs or not. Constraints could arise in several areas: environmental factors, and other constraints and bottlenecks. We briefly consider each.

Environmental Factors

Two environmental factors stand out as potential impediments to oil displacement resulting from coal conversion: stack-scrubber availability and reliability, and the resolution of environmental problems associated with the development of western coal.

Scrubber Problems. Implementation of the conversion program for both the base and extended scenarios will require vastly expanded use of stack scrubbers. Aside from the ever-increasing cost of scrubbers, there is some question of whether scrubbers can be used widely in the near future.

Currently, 25 to 30 firms produce scrubber systems, and FEA estimated in 1975 that their capacity was not then great enough to meet the demand for scrubbers generated by FEA's conversion program. Surveys of the scrubber industry's capacity suggest that it would be possible to meet the forecasted 1980 demand arising both from plants converting to coal and from other plants meeting NSPS or primary standards. However, manpower constraints and extended lead times could raise problems.

FEA estimated scrubber lead times, from design through construction, at about five years. However, the Commerce Technology Advisory Board suggested that, depending on assumptions about the difficulty of acquiring a sludge-disposal site and installing the scrubber, the lead time could stretch from seven to nine years.

A major area of disagreement and uncertainty about scrubbers is their reliability. Though the EPA and other proponents of strict environmental policy are highly in favor of scrubbers, the performance of most actual scrubber installations under conditions appropriate to large-scale introduction has thus far fallen substantially short of reliability targets. The issues do not appear settled. Reliability will improve with greater experience, but progress may be much slower and more costly than FEA assumed in its ESECA benefit-cost analysis.

Other problems include increased particulate-control problems and sludge-disposal problems.

Strip Mining of Western Coal. FEA forecasts of coal production in 1985 and 1990 depend heavily on the expansion of the low-sulfur coal fields in the Great Plains and western mountain states. FEA forecasted that between 1974 and 1980, the rate of growth of coal production west of the Mississippi would be 16.3 percent. From 1980 to 1985, the growth rate is expected to be 8.6 percent, and from 1985 to 1990, 8.7 percent. These very high growth rates may be infeasible.

The Clean Air Act and its interpretation originally greatly stimulated the mining of western coal, with its low-sulfur content. However, a series of environmental constraints was placed on the mining of western coal; in fact, the supply of western coal, which is surface mined, may show a decline for the 1970s because of passed and pending environmental legislation.

The Sierra Club v. Kleppe suit no longer poses an obstacle to mining development. The U.S. Court of Appeals, on June 16, 1975, ruled that no mining development could continue in the northern Great Plains until the Department of the Interior issued a regional environmental-impact statement on coal development. Consequently, an injunction was imposed on four companies, halting their development operations. The Department of the Interior then appealed to the Supreme Court, which granted certiorari on January 12, 1976, lifting the injunction. The Supreme Court's subsequent opinion reversed that of the Court of Appeals.

Production on leased federal land also has been delayed by the need for environmental-impact statements. New leasing of federal lands was halted in 1971 because of an inadequate federal environmental-impact statement. This delay in federal-land leasing has had a significant effect on coal supply because about 40 percent of all coal deposits in the West are on federal land; furthermore, around 85 percent of the recoverable low-sulfur reserves in the United States are under public lands in the West. The Department of the Interior lifted the moratorium in January 1976 and planned to allow bidding on leased tracts that year, with actual leasing beginning in 1977. However, Congress is expected to pass new, stricter legislation on coal leasing that will override the Department of the Interior's recent action.

Land reclamation, which is now required by most states and may be extended by federal legislation, has increased costs and delayed development and production. If revegetation is required in the West, costs will rise because the land in question is either arid or semiarid, and because revegetation will demand large quantities of water.

Unresolved environmental issues have delayed expansion of the coal supply and have increased the costs of production. Whether or not large-scale development of low-sulfur western coal will take place within the next few years is unclear. The delays in the Alaskan pipeline indicate that such litigation and uncertainty can be carried on for several more years.

Transportation, Equipment, and Other Factors

Transportation is potentially a bottleneck for expansion of coal supply in the short run. Although most low-sulfur coal is found in the West, the current markets for this coal are in the eastern and central sections of the country. Therefore, some new rail lines will have to be built, as well as railroad cars, particularly hopper cars for unit trains. Besides rail lines and cars, more barges and possibly pipelines will be needed to transport the coal over long distances. A lead time of at least two or three years may be needed. In the meantime, transport bottlenecks will contribute to increases in the delivered cost of coal.

There is also a potential equipment-availability bottleneck that may affect coal supply. There are shortages in equipment for mining western coal, and the purchase, manufacturing, shipping, and erection of the large stripping machines needed can take up to five years.

The Coal Mine Health and Safety Act also caused some equipment problems. Because the new equipment regulations called for more roof timbers and better ventilation systems, equipment had to be retrofitted on old mines or included in the design for new mines. These alterations increased production costs. Finally, if there is a bottleneck in the supply of scrubbers, mining of high-sulfur coal will not expand.

A variety of other uncertainties could limit coal supply. These factors include a lack of long-term contracts with consumers, and price resistance on the part of some consumers; the scope and timing of nuclear power and the future of oil imports; oil-price regulation, natural gas deregulation, and sulfur-emission regulations.

Another constraint on expanding coal supply is the 30-percent severance tax on western coal production. This severance tax will make western coal less competitive with midwestern coal in midwestern markets. Finally, many of the low-sulfur reserves are captive reserves and are not available to utilities. Thus the feasibility and cost of the coal-conversion scenario depends on a wide range of policies. The coal-conversion option cannot be considered in isolation.

Coal conversion is potentially an extremely cost-effective method of mitigating the adverse effects of a future embargo. As a strong force within the public sector, pressing for removal of obstacles to coal conversion, ESECA can reinforce the attempts of utilities to burn less expensive fuel. Even though FEA and ESECA cannot claim credit for the fact that virtually all new steam-electric capacity ordered after 1975 has been designed for coal-burning capability, the construction orders provide an element of public endorsement that may be of value to the utilities in light of the obstacles from intervenors and public officials oriented exclusively or primarily to environmental concerns.

Coal-conversion policy does not involve large income-redistribution effects, potential problems of vested interests, or the risk of an indefinite continuation of financial commitments. No large federal expenditures are involved. Although technological uncertainties are much fewer than for synthetic fuels, they do pose a significant problem relative to scrubbers during the early 1980s. In addition, environmental problems may be substantial.

It is clear that if coal conversions can be effected, the policy would be cost effective; however, it is not clear that the policy can be fully effected, particularly for higher conversion levels than those of the base case. Electric utilities have resisted conversion both where they cannot obtain assurance that environmental authorities will allow them to burn economic grades of coal, and where the burning of coal would be permitted only in conjunction with scrubbers and other equipment of doubtful reliability. The success of increased coal conversion depends on realistic and consistent environmental-policy administration, in conjunction with improved scrubber technology to control sulfur emissions.

Reduction of Nuclear Lead Times

Reduction of nuclear lead times operates to reduce imports of residual oil principally by lowering the cost of electricity generated with nuclear energy and by lowering the heavy financing burden associated with that cost. The cost reduction results from the effect of the shortened construction period on interest charges and on cost escalation in terms of the investment outlay tied up in the facility under construction. This cost reduction improves nuclear power's competitive position relative to other fuels. In addition, shortening lead times allows oil displacements to occur more rapidly than would otherwise be the case.

The oil-displacement effect of shortened nuclear lead times is more indirect than in the case of coal conversions, and would be modest. The effect of reduced lead times would be to increase the nuclear share of base-load generation, for which virtually no oil will be used; there will be only a moderate and indirect effect in reduced utilization of oil-burning intermediate load and peaking equipment.

The lead-time problem to date has been primarily regulatory and institutional rather than technological or economic. As we discuss below, the minimum presently feasible lead time for nuclear facilities (about seven years) is no less than the actual lead times experienced for nuclear generating facilities begun in the early 1960s, before the licensing process became the dominant pacing element. By the 1970s the lead times had increased to ten years. To reduce lead times to the seven-year minimum consistent with current safety and environmental standards, the licensing process would have to undergo substantial reform. Our procedure was to estimate feasible lead-time compression and the consequent increase in nuclear capacity and oil displacements, and to value the oil displacements and the net costs of the equivalent nuclear-generated electricity.

Policies to Reduce Lead Times

The Nuclear Regulatory Commission (formerly the AEC) has employed a number of procedural and organizational devices, some of them innovative, to expedite the licensing process. For example, environmental, safety, and antitrust hearings are conducted in parallel sessions before separate boards; preapplication filing of information by companies is encouraged to gain a head start in processing cases; regulatory schedules are closely monitored and targeted; and antitrust-license conditions have usually been negotiated. Yet none of these efforts has as yet succeeded in reducing the actual lead times.

The most promising and innovative policy instrument the commission has considered for shortening lead times is the standardization of designs, with preapproval of standard design features. First the AEC, and then its successor, the Nuclear Regulatory Commission, have passed a series of resolutions that encourages standard nuclear-plant design for plants with maximum capacity of 1,300 megawatts, and early site reviews. The objective of these actions is to shorten the licensing-process period. Utilities referencing a standard design will receive priority in the review schedule; additional licensing requirements will be less likely. In addition, there will be likely savings in equipment design. However, in light of the Three Mile Island experience, the futures of either such policies or any acceleration are unclear.

Impediments to Lead-Time Reduction

The NRC's efforts to develop streamlined licensing of standard designs appear to be a well-thought-out and promising method for reducing lead times, but there is as yet no assurance that it will achieve its targeted reduction in NRC processing time. NRC and AEC have a long history of efforts to expedite reviews; unfortunately, the ambitious target dates usually have not been met. Moreover, NRC is frequently not the only regulatory agency to act as the pacing element in determining nuclear lead times; state licensing authorities, often several in each state, usually are involved. Equally important, every hearing is a potential forum for intervenors, some of whom are present because they feel all nuclear-power plant development should be prevented because of generic problems of nuclear safety, theft risks, and radioactive-waste disposal. Finally, there are some limitations or weaknesses of the standardization approach that limit the lengths to which the policy can be pushed, and which could lead to increased costs of nuclear energy in the future. One problem is that implementation of the program could have significant impacts on the relative competitive position of manufacturers of nuclear generating equipment. Another is that the standardization program could develop to the point that the regulatory process discouraged the adoption of innovative, and hence nonstandard, design features. Whether either of these problems becomes serious depends on how the program of standardization is developed and implemented.

Streamlining the NRC licensing process will not, in itself, make state licensing more expeditious. As the NRC reduces its licensing period, state regulation is likely to become the main pacing element, and coordination of federal and state policies will become more important.

Lead-Time Reduction Scenarios

The minimum period for nuclear construction, from the first nuclear-plant concrete to initial operation of the generating facility, is about five and a half years. However, this minimum is achievable only if regulation does not affect the construction period. Since the participation of intervenors in licensing proceedings and in close review of all phases is the result of a conscious and well-established policy to allow public participation in the licensing decision and to assure quality control, it is inconceivable that current lead times of ten to 11 years can be reduced to minimum construction time. With the use of standard nuclear reactors and more uniform standards, the maximum feasible compression in lead times is judged to be seven years. We assume a seven-year period for nuclear construction and licensing. We view this as an ambitious but achievable target.

Nuclear-capacity additions planned through 1984 essentially are locked into completion dates that are no earlier than schedules, though some deferral is possible. Even if steps were taken immediately to implement lead-time reduction policies, the efforts involved in processing the present regulatory caseload would still require considerable time. Consequently, our scenario begins with units scheduled, as of 1976, to begin operation in 1985. We assume that these and all subsequent units will have a seven-year construction period.

Benefit-Cost Evaluations and Policy Considerations

We have assumed that reduction of nuclear lead times involves no net implementation costs, because lead-time reduction, in all probability, would save at least as much in the legal and regulatory resources now expended in protracted disputes as it would incur in program implementation. If the policy succeeds, the benefits in reduced cost of energy would be substantial. The policy is cost effective, quite apart from its effect on oil imports.

To calculate net benefits from oil displacement, it is necessary to estimate the difference in generating cost that would occur from substituting nuclear generation indirectly for oil-fired generation. Without a specific analysis of the electric-power system displacements and costs involved, we cannot employ more than a rough estimate.

Using an interest rate of 10 percent and assuming construction expenditures evenly spread over the lead-time period, we have calculated that reduced lead times would lower the cost of nuclear capacity in 1980 or 1990 from $900 per kilowatt of electricity (kWe) (1975 dollars) in the base case to about $763 per kWe (1975 dollars). At a 70 percent lifetime-capacity factor and a 30-year service life, these figures amount to 2.4 mills per kilowatt hour (kWh) in real costs. In addition, reductions in taxes correlated with capital outlays would further reduce the cost of nuclear energy relative to fossil capacity. These reductions become available after 1985.

However, the incremental nuclear capacity becoming available after 1985 would displace fossil-fueled generation that is initially less expensive than nuclear energy in the base case. We assume a linear interpolation from the 2.4-mill reduction for the first nuclear kWh that does not simply replace base-case nuclear capacity, to the zero reduction for the marginal nuclear kWh. Under this assumption, a percentage differential of 1.2 mills per kWh, half the reduction in nuclear-electricity cost, is the estimated overall cost saving for displacement of fossil-fueled electricity.

To compare this saving in present-value terms with the benefits of other policies is misleading because of the late initial date of benefits for this policy, in conjunction with the 1990 cutoff date employed in the benefit-cost calculations. The relevant point for public policy is that the benefits are substantial, and that the policy involves no necessary costs. Furthermore, the policy avoids disbenefits in that utilities will add the nuclear capacity only when they expect it to be more efficient (less costly) than alternative generating sources. If OPEC collapses in 1980, nuclear construction schedules and capacity plans could be adjusted so that the reduced lead times would impose no net cost.

If the average cost saving per incremental nuclear kWh is assigned to the net value of each oil-fired kWh displaced, then we obtain a value of $1.27 per barrel for direct oil displacements in the OPEC-stability scenario. There is no added benefit in the event of a 1980 embargo, because reduced lead times do not affect nuclear facilities installed before 1980. For the 1990 embargo scenario, the quantity of import displacement is largely unaffected, because the nuclear capacity is already used for base-load generation; on the other hand, the value of the benefit would be increased substantially because of the high price of oil during the embargo.

Like coal conversion and stockpiling, and unlike the other policies considered, lead-time compressions create no new vested interests or inefficient and policy-dependent classes of producers. The impact on the federal budget will be minimal. There need be no redistribution of income between producers and consumers.

High uranium prices and uncertainty over the availability of nuclear fuel may also be limiting factors in the expansion of nuclear capacity, in spite of the potential effect of reduced lead times. Uranium markets are currently in a state of extreme uncertainty. Prices have risen rapidly, the prospects and costs of new reserve discoveries are highly uncertain, and the availability and cost of nuclear enrichment facilities are highly uncertain. At present, all enrichment is performed by the Energy Research and Development Administration (ERDA). Entry by commercial firms may or may not be permitted in time to add capacity by the later 1980s, when ERDA's planned capacity may or may not be adequate to meet the increased nuclear expansion implied by our prospective scenario.

Improved coordination of federal and state environmental and siting policies is needed for the success of this policy. As long as the broader public has not resolved its position on the complex and far-reaching issues affecting the adequacy of existing practices and standards with respect to reactor safety, disposal of radioactive wastes, and protection against nuclear theft or sabotage, litigation and the threat of moratoria or substantial tightening of standards can be expected to impede the rate of introduction of nuclear power.

NOTES

1. These appraisals are discussed at length in Charles River Associates, _Policy Implications of Producer Country Supply Restrictions: The World Energy Market_ (Cambridge, Mass.: CRA, 1976), chap. 6 and app. 4A.

2. See ibid., app. 4A, for assumptions and methods.

5
BAUXITE AND ALUMINUM: IMPACT AND POLICY ANALYSIS

While the United States is even more dependent on a foreign producer association for its bauxite and, hence, aluminum supplies than it is on OPEC for energy, the policy problem with bauxite is substantially different. Embargoes are less probable than in the case of OPEC; however, a sudden cutoff in imports would be very costly. A shift to domestic resources as a result of rising import prices, while costly in total, would have little noticeable effect on most final-product prices. Aluminum is widely used in the economy, but the effects of even substantial increases in bauxite prices would not be disastrous to the country.

OVERALL POLICY CONCLUSIONS FOR BAUXITE

Of the markets considered, bauxite ranks below petroleum and copper in total value of imports; import dependence in bauxite is higher than in these two markets, but lower than for the other commodities considered. As in petroleum, the United States and the rest of the industrialized world face a strong producer association, but the IBA's policy is less fully formed than OPEC's. If the IBA follows the history of past successful cartels, it may maintain its grip on the market for about six years; our market analysis shows that the forces of internal and external competition that have destroyed past cartels may be weak in the case of bauxite, and that the IBA may maintain its hold on the market for some time.

If indeed the IBA will remain effective for some time, U.S. policy should have two primary purposes: to guard against the possibility of extreme short-run embargoes or pricing actions, and to moderate long-run IBA policies (we assume policy aims to further

U.S. economic interests and abstract in general from broader social and diplomatic purposes). The first purpose would be well served by a stockpile equivalent to perhaps six months to a year of consumption. It is conceivable that political events in the Caribbean could motivate an OPEC-type embargo or an individual country cutoff, the costs of which could be very high. The stockpile hedge might also be strengthened by an announced U.S. commitment to deny future imports (perhaps only for a fixed time) from countries engaging in bauxite embargoes, though this would carry its own risks and costs.

For the longer run, the United States has a number of options. On the nontechnological side, it is clearly in U.S. interest to induce the still-growing producers, Australia and Brazil, to continue expansion even at the possible sacrifice of some of their narrow economic interests in the bauxite market. Possible U.S. levers include diplomatic pressure or encouragement combined with trade incentives, preferences, and concessions. Australian and Brazilian moderation would not only force other IBA countries to practice moderation, but could increase competition for market shares sufficiently to break the cartel.

Other long-run measures include locating and developing alternative bauxite sources, and improving the cost and credibility factors of the domestic bauxite and nonbauxitic alternatives. Research to improve the energy efficiency of clay-conversion processes, which produce aluminum from domestic resources, and the construction of commercial-scale demonstration plants could have these effects. Bauxite is prototypical of materials which are available or replaceable at some, though not extreme, additional cost from domestic sources, and which constitute a small proportion of final-product cost. For such commodities, supply-side policies hold greater promise than efforts to encourage conservation or to increase the ease of substitution.

The Bauxite and Aluminum Markets: Limits on IBA Power

Valued at approximately $600 million in 1974, net U.S. imports of aluminum, alumina, and bauxite are quite small relative to crude oil imports, which were over $15 billion in 1973. The aluminum market is substantial, with 1973 consumption of about 6.8 million short tons, for a total value of about $4.7 billion. The market is clearly a vital one; although substitutes exist for aluminum in most of its end uses, the overall price elasticity of demand of about -0.1 in the short run, and about -0.7 in the long run, is moderate, which indicates that there are substantial advantages to users from using aluminum rather than its substitutes. A shortfall in aluminum supply would have major dislocative effects on the using markets, as could major increases in the price of aluminum.

The IBA has a partial monopoly with respect to low-cost sources of bauxite, although not with respect to aluminum. For example, at 1976 Jamaican tax levels, the cost of Jamaican bauxite constituted only 12 percent of the 1976 cost of aluminum made using it. The increase in Jamaican and other bauxite taxes, from about $1.80 to $15.00 per ton by 1976, approximately doubled the delivered cost of bauxite in the United States, and raised the cost of producing aluminum by about 6 percent; if sustained, these actions would increase aluminum prices by this proportion.

A doubling of Jamaican and all IBA taxes to about $30 per ton, in 1971 dollars, is the maximum which could be sustained in the long run because of the cost ceiling imposed by domestic bauxite and the essentially unlimited domestic deposits of nonbauxitic ores. This maximum sustainable cost increase would further increase the cost of producing aluminum by about 6 percent. Applying the long-run demand elasticity, this price increase would, over several years, lower total consumption by approximately 4.3 percent below what it otherwise would have been. Over the very long run, the decrease would probably be somewhat larger. The actual effect of the price increase will be gradually to slow the growth rate of aluminum consumption, not suddenly to depress its level. Therefore, there need be no significant effects on employment or output in aluminum-producing or -using industries.

Short-run IBA actions such as embargoes or extreme short-run monopolistic pricing appear very unlikely; the Caribbean producers have, in relative terms, a greater stake in the U.S. market than the United States has in access to Caribbean bauxite. In the case of such actions, the United States could rely on stocks in the short run (if they were available) and shift, over a period of years, to domestic alternatives.

Since sustained and large aluminum price increases are unlikely, and substitution possibilities generally are well known, demand-side policy options are less important than supply-side alternatives. That is, the total demand for aluminum will be relatively unaffected in response to actions by foreign bauxite producers. The long-run policy questions, then, involve the availability and cost of supply-side alternatives and the possible effects of government policies. The short-run policy questions include the possible impact of extreme IBA actions and the usefulness of stockpiling and other policy options.

The Initial IBA Actions: A Crisis?

The IBA tax escalation, which may have cost the United States as much as $160 million in 1975, did not precipitate a crisis in the bauxite market. Supplies remained available and aluminum production and use were not disrupted. Jamaica and other Caribbean producers had for some time been interested in forming an export cartel, and rumors of its impending formation had surfaced several times in the past. Given the Caribbean countries' cost advantage in the U.S. market and Jamaica's leading position and relatively low costs, it is not surprising that Jamaica led the tax increases.

OPEC's success in raising oil prices appears to have stimulated the IBA's formation in two ways. First, it provided an example of cartel success. Second, the oil-price increases imposed enormous balance-of-payment burdens on the major bauxite-producing countries. Jamaican oil costs tripled from $55 million in 1972 to $165 million in 1974, compounding existing balance-of-payments problems. Inflation in the United States and Europe, caused in part by higher oil prices, had raised the costs of imports and contributed to Jamaican financial troubles.

RESPONSES TO IBA TAX INCREASES

U.S. Government and Producer Reactions

The depressed state of the aluminum market in 1974 and 1975 and government opposition prevented producers from fully passing on the bauxite-price increases in the form of higher aluminum prices. A substantial part of the increase has come out of the aluminum companies' profits. When Jamaica broke off negotiations with the companies and unilaterally increased taxes, Alcoa, Kaiser, and Reynolds brought suit before the International Center for the Settlement of Investment Disputes (ICSID), and Alcoa also appealed the new taxes under provisions of Jamaican law. The aluminum companies also requested U.S. diplomatic support in reversing the Caribbean tax increases, but expressed disappointment with the results.

Under U.S. law, tariffs and other penalties exist for violations of ICSID decisions, and these penalties probably restrain producer actions. On the other hand, the aluminum companies' large investments in producing countries could be endangered by very active public or private actions to counteract tax increases. Conversely, loss of the U.S. market due to extreme pricing or other actions could cost IBA members the entire U.S. market; as of 1975, the current value of this market to the IBA members for the 1975-90 period was several billion dollars.

Since the past tax increases did not precipitate a crisis, we have not analyzed possible past contingency policies in detail. As discussed below, the costs of domestic alternatives are high enough so that subsidies for domestic bauxite and nonbauxite-ore production would have imposed much larger costs than benefits. Also, because reasonable tariffs would cause only very small decreases in bauxite consumption, a tariff policy would not seem appropriate unless IBA taxes rise to the limit set by the cost of processing domestic nonbauxitic ores.

Some stockpiling releases would have been economically beneficial: the Jamaican tax increase of about $13.20 is the maximum potential initial per-ton gross benefit that stockpiling could have provided against the price increase. The actual increase was less since other IBA countries did not fully or promptly follow the Jamaican increase. Table 5.1 shows the costs, at the end of 1974, of stockpiled bauxite acquired at a price of $13.19 per ton during various years before 1974, assuming total storage and interest costs of 10 percent per year.

TABLE 5.1

End-of-1974 Cost per Ton of Stockpiled Bauxite Acquired at $13.19 per Ton

Year Acquired	Cost per Ton of Stockpiled Bauxite, End of 1974 (initial cost plus accumulated interest and storage cost)
1970	$21.23
1971	19.30
1972	17.54
1973	15.95
1974	14.50
Mean	17.70

Source: Charles River Associates, Policy Implications of Producer Country Supply Restrictions: Overview and Summary (Cambridge, Mass.: CRA, 1976), p. 113.

The per-ton initial benefit of stockpile releases in 1975 was about $26.40, the delivered price of Jamaican bauxite. The initial costs shown in Table 5.1 indicate that if the price increase had been expected with certainty, stockpile releases could have yielded considerable benefits; since the demand for bauxite is highly insensitive to price, an efficient release rate could substantially have displaced imports. The maximum efficient release for 1975 would have been between 20.16 million tons, totally displacing imports, and 6.48 million tons, displacing Jamaican imports. The higher release rate assumes that Jamaica's tax increase applied to all imports. The lower end of the range assumes that only Jamaica's price increased, so that the correct optimal release is between these values.

If the price increase had been uncertain, but had been expected with a probability of .5, then the expected gross benefit from initial stockpile releases in 1975 would have been about $19.80 for a per-unit expected net benefit of $2.10 (assuming that stocks had been acquired at an even rate over 1970-74). If releases had displaced all Jamaican imports, including bauxite contained in alumina, then the total expected net benefit in 1975 would have been $13.6 million ($2.10 times 6.45 million tons); displacing all imports would have yielded total expected net 1975 benefits of about $42.3 million (which is $2.10 times 20.16 million tons). If the expected stockpile releases had entirely deterred the price increase, the gross benefits would have been as much as $295 million, assuming all supplying countries had matched Jamaican taxes; if the presence of the stockpile had permanently deterred the price increase, the present value of the gross savings until 1990, at a 10 percent real rate of discount, would be as much as $3 billion; holding a stock of one year's imports for the assumed 20 years would cost on the order of $1.8 billion, for a net maximum deterrence benefit of about $1.3 billion (in 1975 dollars). However, it is very unlikely that a one-year displacement of imports would permanently deter the IBA action. These levels of stocks and stockpile releases were within the range of U.S. government stocks in excess of stockpile objectives prior to October 1976, which could have supported approximately one year of aluminum consumption. However, stockpile goals were raised in October 1976, so that a deficit in excess of a half-year of U.S. consumption was created.

The available inventory data do not present evidence of private-sector anticipation of the price increase: the private inventory-to-consumption ratio was lower in 1974 than in 1970. The private sector appeared to believe that the likelihood of a price increase did not justify substantially increased stocks; it is also possible that the industry could have believed that, in the event of a price increase, government stocks would become available.

Private-Sector Technological Preparation for Possible Crises in the Bauxite Market

However motivated, private firms have acquired domestic bauxitic and nonbauxitic ore properties, produced some bauxite domestically, and developed processes for the use of nonbauxitic ores. Details of these processes and their costs remain proprietary. For example, National Southwire is testing a pilot plant for producing alumina from alunite, which may prove commercially viable even at present bauxite prices. Anaconda has shown interest in clays, particularly Georgia kaolin; Alcoa has investigated anorthosite and coal tailings as aluminum sources; and Pechiney is building a commercial-scale plant using a shale-based process to produce aluminum.

Various motivations may underlie these efforts, not least among them being normal commercial interest in lowering costs and diversifying supplies. The aluminum companies may have feared future price increases or difficulties in imported bauxite supply, or wished to strengthen their position in bargaining with producing countries.

As discussed below, processes to use domestic aluminum sources are known and, for domestic bauxite, are currently in use; however motivated, private-sector actions have done at least part of the job of contingency planning for potential future crises. Severe problems appear unlikely, so the preparations are, in all probability, adequate.

LONG-RUN SCENARIOS OF IBA BEHAVIOR

This section compares the impacts of two long-run pricing scenarios against a base case, and considers a variety of policy options available to the government to deal with supply restrictions. Policies considered include strategic use of a tariff policy, stockpiling, a variety of technological options including some affecting the costs of using domestic resources, and a range of diplomatic and other noneconomic measures.

To evaluate long-run options, we compared the following three cases (all values are in 1976 dollars):

1. Base case: IBA taxes remain constant from 1975 to 1990 at their pre-1974 level of slightly below $2.00 per long ton of bauxite.
2. Moderate IBA policy: In 1977, Australian and Brazilian taxes remain low, and the taxes levied by the other IBA countries fall to an average of $9 per ton.

3. IBA limit pricing: In 1977, Brazil and Australia match IBA prices, with a $2 per ton differential for Brazil to allow expansion; Caribbean taxes rise from $15 to $20 immediately, and in 1985 rise to $26 per ton. Long-run IBA prices are limited by the costs of processing domestic ores.

Another possible scenario is an IBA collapse due to the discovery of new deposits or a disagreement over appropriate market shares. Even without new discoveries, it appears that the second, or moderate, price policy is more likely than the limit-pricing case, which requires Australian cooperation in a high-price policy. If the IBA did collapse, taxes probably would revert to a level between the pre-1974 level and the moderate-price scenario level.

Impacts and Costs of Alternative Long-Run Scenarios

While both the moderate- and limit-price scenarios impose substantial costs on the United States and other consuming nations, they do not result in major disruptions in the market.

Under the base (low-tax) case, U.S. aluminum consumption increases by 94 percent between 1978 and 1990; even under the limit-price policy, it grows by 62 percent.

By 1990 we estimate that the moderate-price policy will have raised the aluminum price by about 2 percent, while the limit-price policy will raise the price by about 8 percent (in real terms). Even under the limit-price scenario, U.S. aluminum consumption is never more than 10 percent below the base-case forecast. This indicates that the two likely scenarios of IBA behavior will not have highly disruptive effects on aluminum users or cause substantial increases in final-product prices. The total costs to the nation are substantial, however, as discussed below.

Indeed, while aluminum prices and use are not greatly affected, the moderate-price and limit-price scenarios greatly increase the cost of bauxite imports and so impose large costs on the country. Table 5.2 shows the estimated costs to the United States of the alternative scenarios, both in current value in 1977 and in 1980 and 1990, as well as total increases in IBA earnings from all importing countries. These values are very substantial, both in total current value and in typical future years.

The total cost to the United States of the two IBA pricing policies includes both increased payments to the IBA, and increased costs and decreased profits to users due to lowered aluminum consumption. We assume that neither price policy results in increased domestic production of bauxite or use of nonbauxitic ores. Because of the small effect on aluminum prices, over 99 percent of the cost in any year consists of increased payments to the IBA. Actual cost to the United States may be higher if sustained high IBA prices result in large expenditures to reduce the costs of alternatives to IBA bauxite.

TABLE 5.2

Effects of IBA Actions, 1978-90
(millions of 1976 dollars)

Effect	Moderate-Price Policy	Limit-Price Policy
Cost to the United States		
Present value in 1977*	800	2,700
1980	107	318
1985	131	542
1990	165	676
Total increase in IBA earnings from all sales		
Present value in 1977*	2,300	7,500
1980	289	862
1985	389	1,613
1990	520	2,146

*Discounted back to 1977 at 10 percent compounded annually.

Note: All magnitudes are relative to base case; IBA taxes, at approximately $1.80 in real terms.

Source: Charles River Associates, Policy Implications of Producer Country Supply Restrictions: Overview and Summary (Cambridge, Mass.: CRA, 1976), p. 117.

The cumulated present value of the cost to the United States is $1.9 billion higher under the limit-price policy than under the moderate-price policy, an increase of a factor of more than two. Thus, the United States has a substantial economic interest in encouraging IBA moderation. A substantial investment might be justified if policy makers were reasonably certain it would induce the IBA to follow a moderate policy. In addition, substantial gains could be reaped by destabilizing the IBA and forcing taxes down to their pre-1974 levels. Table 5.2 indicates that if success were assured, present investments of between one and two billion dollars might be justified; a 25-percent chance of inducing a return to the base case would justify an investment of several hundred million dollars.

The IBA has a great deal to gain from cartel stability and movement toward higher prices. Raising taxes from the base level yields a present-value gain to the IBA of over $2 billion; movement toward the limit price yields over $5 billion more. While the actual gain from a move to limit pricing may be less than we have estimated due to very long-run substitution and competitive pressures on prices, it is clear that IBA countries have a very substantial economic interest in maintaining stability in the association and moving to high prices. It is also clear that the United States and other consuming nations have a very substantial interest in moderating IBA behavior.

Before discussing policy options, we should note that we consider the limit-price scenario and its associated high costs highly unlikely. Its success requires Australian cooperation and an IBA belief that price-induced development of new producing areas are unlikely. In addition, this limit-price policy pushes the cost of IBA bauxite close to that of domestic alternatives, which are available in essentially unlimited supply. The costs of these alternatives could fall, due to technological change, commercial experience, or lower energy prices. If so, the IBA would be forced to moderate its price policy or risk losing the U.S. market. While the presence of large capital investments in IBA countries could prevent the aluminum companies from shifting quickly, over the longer run, IBA power and earnings would decline.

Policies to Moderate IBA Behavior and Lessen the Impact of Cartel Pricing

Assuming stability (and no lower priced imported alternatives), the long-run limit on IBA taxes, and hence bauxite prices, is set by the cost to users of domestic alternatives (the aluminum companies). Our initial judgment that the IBA may be stable for some time is based on the concentration of known low-cost reserves and resources, the ease of detecting cheating on cartel prices, and the IBA countries' strong collective and individual interest in high prices. Policies may have impact in these two areas by lowering the ceiling on bauxite prices and by decreasing stability.

We now consider two types of policy options: first, policies designed to prevent movement toward higher prices, as epitomized in the limit-price scenario. Such policies include technological development, to lower the ceiling price and increase domestic supplies; tariffs and quotas; and diplomatic policies. The second type of policy is stockpiling as a hedge against the possibility of movement to the limit-pricing policy.

Technological Policies

Technological policies include lowering the cost and enhancing the credibility of domestic alternatives, improvements in bauxite prospecting and deposit evaluation, and lowering the cost of transporting bauxite and aluminum.

Lowering the Cost and Enhancing the Credibility of Domestic Alternatives. Processes for using domestic nonbauxitic ores are highly energy intensive, and improving heat recovery would substantially lower costs. For example, according to our calculations based on data from the Bureau of Mines and on a recent study performed by Kaiser Engineers, the hydrochloric acid-ion exchange process is at present the least expensive method for obtaining aluminum from domestic nonbauxitic sources. Two alternative crystallization methods have been proposed for the hydrochloric acid process—evaporative and gas-induced crystallization. The second, developed by Kaiser Engineers, has not yet been tested experimentally, but offers large potential energy savings that may make the method competitive with bauxite at current prices. Presented in Table 5.3 is the approximate price, per metric ton, for bauxite at which alternative nonbauxitic aluminum processes are economical.

Reducing the cost of using domestic bauxitic and nonbauxitic ores could lower substantially the ceiling on IBA prices, reducing the likelihood of the high-price policy and decreasing its potential impact. For example, at projected import rates, a $1 decrease in the ceiling would reduce the cost of the limit-pricing policy, should it occur, by about $30 million per year by 1985. It thus appears that significant private and/or public projects to reduce the cost of domestic alternatives could have economic justification, depending on the prospect of success. If such prospects moderated IBA behavior, they would confer benefits not only on the firms involved in the alternative technologies, but also on all aluminum users.

Continued pilot-plant testing of methods to refine nonbauxitic ores could also be desirable, particularly if done in near-commercial-scale plants. Present testing is in the miniplant stage, and commercial-scale operations could set more meaningful benchmarks and act as a deterrent, particularly if the costs were lower than previous estimates and the results were conveyed clearly to the IBA members.

TABLE 5.3

Approximate Price per Metric Ton for Bauxite (5 Percent Moisture) at Which Alternative Nonbauxitic Aluminum Processes Are Economical (in 1976 dollars)

Process	Price per Metric Ton
Anorthosite	61
Clay-hydrochloric acid-ion exchange	
Evaporative crystallization	44
Gas-induced crystallization	28.7

Source: Charles River Associates, Policy Implications of Producer Country Supply Restrictions: The World Aluminum-Bauxite Market (Cambridge, Mass.: CRA, 1977).

We should note that the cost estimates presented in Table 5.3 are tentative and may be biased upward, as may be the costs of the IBA limit-price policy derived from them. While the cost disadvantages of the nonbauxitic processes have risen in recent years, the costs of the newer processes, such as the hydrochloric acid-ion exchange, are as much as 50 percent below the costs of older processes. It thus appears that research into methods of reducing process costs could yield substantial benefits, especially if policy makers find the risk of IBA cohesiveness is substantial, or wish to hedge against the possibility of extreme short-run IBA action.

Improvements in Bauxite Prospecting and Deposit Evaluation. Various remote tropical and subtropical areas appear to have appropriate geologic characteristics to contain bauxite deposits. The discovery and development of new deposits could be accelerated by research into better methods for prospecting and for distinguishing economic from subeconomic deposits. If based on research done by the government, such techniques could be made available to all potential bauxite developers. The discovery and development of substantial new high-grade deposits could not only prevent adoption of the limit-price policy, but also weaken IBA stability and force taxes toward the base-case level. The benefits to the United States in such a case could be in the billions.

Lowering the Cost of Transporting Bauxite and Aluminum. Transport-cost reductions for bauxite and aluminum would reduce the relative cost of using ores from potential deposits in remote areas; they could also lower the cost of using Australian sources, which is particularly important considering Australia's likely low-to-moderate tax policy. A reduction in inland-transportation costs, both in the United States and abroad, could have similar effects in reducing the Caribbean cost advantage and undermining IBA stability.

Tariff, Quota, and Subsidy Policies

The appropriate tariff policy could constrain IBA pricing power and lower the probability of limit pricing or other high-price policies. As the cost of obtaining aluminum from IBA bauxite approaches the cost of domestic alternatives, the United States has less and less to lose by imposing a tariff or quota high enough to induce switching. The appropriate tariff or quota could force the IBA to reduce its taxes so that its delivered price, including taxes, would be below the cost of domestic alternatives.

Such a tariff policy would allow the U.S. government to capture some of the IBA's potential profits and would impose heavy costs on the IBA. Jamaica and other Caribbean producers would lose substantially, both directly and indirectly, during the struggle which would ensue to gain increased shares of the non-U.S. market; cartel stability would be decreased. It thus appears that the U.S. bargaining position could be quite strong under a tariff or quota policy, although the practicality of the policy is limited by the income redistribution from aluminum producers and users to the U.S. treasury as well as political considerations.

Since bauxite demand is highly inelastic, the cost to users due to decreased use would be low, although the tariff would redistribute income from aluminum consumers to the U.S. treasury. While such revenues could be used to compensate consumers and producers, the affected groups would be likely to object strongly.

In addition, if the tariff policy induced aluminum companies to consider shifting to domestic or other sources, the IBA countries might move to expropriate U.S. investments in mines and other facilities. A threatened tariff might cause the IBA to mount an embargo to deter or moderate the tariff policy. A further potential problem is that tariffs on bauxite and alumina alone could induce offsetting increases in aluminum metal imports. While this problem could be solved by a tariff based on the IBA bauxite content of imported aluminum, such actions would be difficult to enforce and could cause highly complex treaty and diplomatic problems, especially with industrialized aluminum-exporting countries.

In some ways a quota might be preferable to a tariff. Under a quota arrangement, import rights could be auctioned, and if excess capacity developed in bauxite or alumina, producing countries might compete to export to the U.S. market. Such a policy would probably be less effective in this market than in chromite, manganese, or possibly the platinum group, as domestic firms are vertically integrated back to bauxite mining. Producing countries have had relatively small roles in the actual marketing of their output. Producer-country movements to downstream integration and increased control could increase their direct involvement in marketing decisions and increase the potential usefulness of the quota option. As with the tariff, there is risk of expropriation or other precipitous producer-country reactions to such policy moves. One element essential for the success of these policy options is an effective demonstration of U.S. determination to enforce the policy, if implemented. If producing countries expect a tariff or quota to be temporary or used only as a threat, its impact is likely to be minimal. In addition, if domestic producers expect the policy to be temporary, they will hesitate to invest in alternative processes.

An announced policy of subsidizing domestic aluminum-ore production and processing could place a ceiling on IBA prices equal to domestic cost less the subsidy. Such a policy would involve direct transfers from government expenditures to domestic producers and would not yield significant output unless the subsidy were large and expected to be permanent. Domestic subsidization also would be politically difficult to implement.

Diplomatic and Other Long-Run Policy Options

Diplomatic options could have the effect of decreasing the likelihood of the limit-pricing scenario by decreasing IBA stability. While many options are potentially available, we consider two: encouragement of Australian and Brazilian moderation in bauxite-export policies, and diplomatic support to companies engaged in bauxite exploration, development, and production.

Encouragement of Australian and Brazilian Moderation. If Australia and/or Brazil adopted moderate- or low-tax policies and expanded exports to a rate consistent with their resources, most of the other bauxite producers would in a few years be forced to adopt similar tax policies. As Table 5.2 indicates, the benefit to the United States of a moderate- as opposed to a high-tax policy is on the order of several hundred million dollars per year. The U.S. government can encourage this outcome by supporting Australia and Brazil in regard to their larger political and economic concerns, and by encouraging the two countries to sacrifice, in part, their narrow economic interests in the bauxite market. In Australia's case it might be useful to give more sympathetic attention to requests for freer access to U.S. markets, particularly for agricultural products. Sympathetic treatment of Brazilian requests for credits or tariff preferences could have similar effects.

Diplomatic Support for Companies Engaged in Bauxite Operations. The United States could offer support to companies engaged in bauxite exploration and assist them in dealing with host-country governments. Once new deposits were identified, the United States could speed development by helping to arrange or offering financial support, particularly for transportation networks and other costly components of infrastructure.

In addition, the United States could support aluminum-company efforts in the ICSID and otherwise encourage moderation on the part of the IBA countries. Such actions should, of course, be highly discreet, but clear statements of the intention to respond firmly to extreme pricing actions could have salutary effects.

Stockpiling Policy as a Hedge Against High Long-Run IBA Prices

As discussed above, the costs of an IBA move to higher prices in the long run (the limit-price policy) might be substantial (depending on the costs of domestic alternatives), though the policy is unlikely. If prices do rise substantially, they are unlikely to do so before 1980 as, prior to that time, expansion in Brazil should constrain IBA power, even if Australia is relatively cooperative with the association. Based on current and estimated future bauxite costs and estimated 1980 aluminum costs, we found that a probability of limit pricing of at least 60 percent is required to justify holding significant stocks from 1977 through 1980 as a hedge against this possibility. The limit-price policy does not appear to be 60 percent likely as it depends on an unlikely degree of Australian cooperation; therefore, substantial stockpiling does not appear desirable as a hedge against higher long-run bauxite prices. However, if significant stocks did themselves reduce the likelihood of higher prices or undermine IBA stability, stockpiling would be a more desirable policy option for dealing with a limit-price policy than it appears to be from this analysis.

IBA EMBARGOES AND SHORT-RUN MONOPOLY PRICING: IMPACTS AND POLICIES

Embargoes and short-run monopoly pricing appear very unlikely because of the harm exporting countries would inflict on themselves by using them, especially if the U.S. government acted to prevent resumption of imports following such an action. Embargoes could arise out of an IBA attempt to force removal of a U.S. tariff to prevent the aluminum companies from shifting to non-IBA sources, or out of an attempt to pressure the aluminum companies into paying higher taxes. IBA countries might also massively raise taxes to take advantage of the time lags and costs involved in shifting to alternative supply sources. We first consider the impact of embargoes in light of available U.S. alternatives, and then consider short-run monopoly pricing.

Impacts of an IBA Embargo

Even under extreme circumstances, it is unlikely that Australia would participate in what would be a hostile act directed against a military ally. The worst possible situation appears to be a total cutoff of supplies from all other major producers, namely, all the other current IBA members plus Brazil and the Cameroons. Including the amount of bauxite contained in alumina, the result would be a loss of about 60 percent of U.S. bauxite supplies, equivalent to about 3.3 million tons of contained aluminum. Unless this shortfall could properly be offset, aluminum prices would rise sharply and usage would decline, and workers and other resources in U.S. producing and using industries could be idled. Users would attempt to shift to substitutes, including copper, steel, and plastics; substitutes are available for all the major uses, though substantial costs would be incurred. Substitutions would be greatest in containers and some transportation uses, and the relative cost impacts would be largest in electrical and some consumer-durable end uses. Since the likelihood of severe and lasting cutoffs or price increases appears small, we do not consider demand-side alternatives further.[1] We turn now to alternative sources of aluminum, including increased bauxite production in the United States and other non-IBA countries, U.S. government stocks, and increased secondary-supply recovery. In the long run, the deficit could, of course, be made up by production of domestic nonbauxitic ores.

Production from Non-IBA Countries and Nonbauxitic Ores

As most of U.S. production comes from Arkansas open-pit mines, operating on a one-shift basis, domestic output could be expanded, perhaps by a factor of as much as three, by adding extra shifts; at higher costs, Arkansas output could perhaps be raised somewhat further by adding surface-mining equipment. Because even a very large increase in the cost of mining bauxite would have a relatively small effect on aluminum prices, even a substantial increase in costs would be tolerable. A fourfold increase in mine output is consistent with the nation's experience during World War II: between 1941 and 1942, domestic bauxite mine shipments increased from 897,000 tons to 2.48 million tons, a factor of about 2.8; and between 1942 and 1943, by a further factor of over 2.0. Much of the increase in mine output thus occurred within a year. Some alumina capacity would be lost due to the low efficiency of plants when recovering domestic ore. The shortfall for the first two years would be about 45 percent; within two years, the alumina capacity needed to increase process efficiency could be made available on a crash basis.

The need for additional mining and alumina capacity points to a potential policy problem: if private industry expects the crisis to be short-lived, it may hesitate to make the needed investments in capacity, in which case some form of government financial assistance, such as loan guarantees or financing and operation of plants by the government, might be economically justified. Given the importance of bauxite and alumina exports to the economies of the Caribbean countries, embargoes not only are unlikely, but almost certainly would be very short-lived if they did occur. Massive domestic investments therefore are not likely to be needed.

Production of alumina from nonbauxitic ores can plausibly be assumed to begin within two years of the onset of the embargo, assuming crash investment efforts begin immediately after the embargo is announced. Three years later, such production, as well as bauxite imports from nonembargoing countries, could increase sufficiently to offset fully the remaining 30-to-45 percent shortfall in production.

The assumptions presented here are conservative, given the possibility of increased Australian shipments to the United States. In addition, we have ruled out the possibility of increased imports of primary aluminum and assumed that the IBA is totally successful in policing the embargo; these assumptions are unlikely to be true, but are used to indicate the maximum possible impact of a crisis in the bauxite market.

Increased Secondary Supply

Increased secondary-supply recovery could make only a very small contribution to the shortfall. CRA's econometric analysis indicates that secondary supply is very insensitive to price. Scrap from obsolete aluminum-bearing products that, unlike new scrap generated in the production process, represents a short-run net addition to total supply accounts for only about 3.5 percent of consumption. Doubling this amount would require approximately a quadrupling of the aluminum price. Given the availability of industry and some government stocks, such price increases are highly unlikely.

It is possible that technological innovation could lower the costs of secondary recovery, but given the potential contribution of secondary recovery, the likely gain appears small.

Cost Impacts of an IBA Embargo

The cost impact of an IBA embargo depends on the embargo's duration and the availability of stocks. The cost of the embargo consists of the extra cost of output from U.S. sources, the value of idled resources due to decreased aluminum output, and the cost to users of decreased use of aluminum. The actual required decrease in consumption depends on the availability of stocks. If stocks are allocated to minimize the reduction in consumption in any given year, the average required reduction in aluminum consumption, which of course determines the resource loss, is as shown in Table 5.4. These results show the potential importance of stocks as a hedge against an embargo; without stocks, a one-year embargo would reduce aluminum consumption by as much as 45 percent, causing substantial economic losses and unemployment. With stocks equal to a year's consumption, an embargo expected to last two years need not reduce consumption at all. For embargoes of plausible length (a year or less), current strategic stocks (if available) could allow consumption to continue unabated. This does not mean that an embargo will have no adverse effects. Rather, it indicates that these effects need not be disastrous and can be controlled by proper stockpile policy. An embargo could affect aluminum-company and aluminum-user expectations and initiate long-run shifts to alternative sources of bauxite supply and materials.

Using our estimates for 1980 U.S. aluminum consumption and prices, and assuming that bauxite stocks are maintained at a level equal to 90 percent of 1980 consumption, we obtain the costs in Table 5.5 for a 1980 embargo of varying durations. These cost estimates are substantial, but are small relative to those which could occur from potential future crises in the petroleum market. Assuming the availability of stocks, the impact of a brief embargo need not be large. It is important to note that current government stocks are sizable, but may not be available during an economic emergency. The GSA stockpile is designated a strategic stockpile for protection against cutoffs resulting from military emergencies.

TABLE 5.4

Average Annual Required Percentage Decreases in U.S. Aluminum Consumption Resulting from an IBA Embargo (embargo duration in years)

Ratio of Stocks to Annual U.S. Consumption	Duration of Embargo (years)			
	1	2	3	4
1.5	0	0	0	0
1.0	0	0	5.0	10.00
0.75	0	7.5	13.3	16.25
0.5	0	20.0	21.7	22.50
0.25	40	32.5	30.0	28.80

Notes: Nonbauxitic production is assumed not to occur for embargoes shorter than three years. Stocks are allocated to minimize the reduction in consumption in any given year.

Source: Charles River Associates, Policy Implications of Producer Country Supply Restrictions: The World Aluminum-Bauxite Market (Cambridge, Mass.: CRA, 1977).

TABLE 5.5

Costs to the United States of 1980 Embargo (current value as of 1980 in billions of 1976 dollars)

Cost Element	Duration of Embargo (years)			
	1	2	3	4
Loss to consumers plus idled resources	0*	1.65	3.56	4.68
Increased domestic production costs	.04	.08	.17	.31
Total resource cost	.04	1.73	3.73	4.99

*Estimate is zero because available stocks are more than sufficient to cover the implied consumption shortfall.

Note: Estimates assume stocks equal to 0.9 years of 1980 U.S. consumption, released in an efficient manner. The length of the embargo is assumed to be known with certainty.

Source: Charles River Associates, Policy Implications of Producer Country Supply Restrictions: Overview and Summary (Cambridge, Mass.: CRA, 1976), p. 129.

Our cost estimates assume that stocks are made available and allocated in a reasonably efficient fashion. If stocks are unavailable, the short-run cost of an embargo could be multiplied manyfold, the consequences of which would be partly mitigated by increased imports of raw aluminum metal and products. In addition, without stocks, the aluminum companies would be likely to bid very high prices for bauxite supplies, putting the IBA countries under extreme pressure to act competitively. If the total probability of a 1980 embargo were 20 percent, which appears very high, and if durations of one through four years were equally likely at 5 percent, then the expected total cost would be about \$912 million, and the expected current-value cost as of the beginning of 1978 would be about \$764 million, discounted at 10 percent. This estimate is almost certainly high; both the overall embargo probability and the relative likelihood of the longer embargoes are unrealistically large.

Stockpiling as a Hedge against an IBA Embargo

A variety of policies might be adopted in anticipation of an IBA embargo, including those considered above as long-term alternatives. In particular, measures to speed the availability of increased domestic bauxite and nonbauxitic aluminum could lower the cost impact of an embargo, as could government construction, operation, or subsidization of standby domestic bauxite, nonbauxitic-ore, and alumina facilities. The likelihood of an embargo seems so small that such programs would not appear to yield net benefits. On the other hand, it is apparent from Table 5.4 that significant domestic stocks can reduce the required shortfall in consumption due to an embargo, and hence substantially reduce an embargo's cost impact. The efficient stockpile levels for a three-year embargo commencing in 1980 for a range of embargo probabilities are shown in Table 5.6.

The present government-strategic bauxite stockpile is equivalent to about 66 percent of forecasted 1980 annual U.S. consumption. A 6 percent embargo probability would be required to justify maintaining a stockpile equal to this size for economic contingency purposes. However, if holding large stockpiles reduces embargo probabilities, then holding larger stocks could be justified. The importance of deterrence effects in determining the efficient stockpile size is discussed further in the Appendix to this volume. A larger stockpile would also be valuable in the event of IBA long-run limit pricing or of a massive short-run increase in IBA taxes (a possibility that we discuss below). A large stockpile should also serve as a deterrent to the latter. Consequently, the optimal economic contingency-stockpile policy may be to hold a stockpile equal to at least six months of U.S. consumption and perhaps as much as one year's consumption.

TABLE 5.6

Efficient Stockpile Levels for a Three-Year Embargo Commencing in 1980

Probability of Embargo	Efficient Stockpile as a Proportion of Forecasted 1980 Consumption
0.22	1.30
0.12	1.16
0.06	.66
0.04	.41
0.03	.22
0.025	.08
0.02	0

Note: The efficient stockpile is the stockpile level that, if deployed correctly, results in the maximum reduction in embargo-cost impacts as shown in Table 5.5, less stockpile costs.

Source: Charles River Associates, Policy Implications of Producer Country Supply Restrictions: Overview and Summary (Cambridge, Mass.: CRA, 1976), p. 131.

Short-Run IBA Monopoly Pricing: Impact and Policies

Either the IBA as a group or individual IBA members might engage in restrictive actions that could not be maintained profitably in the long run. Consideration of all possible cases is beyond the scope of this project, but below, we briefly consider two prototypical cases: attempts by individual countries or groups of countries to take advantage of the lack of consuming countries' short-run alternatives; and concerted increases in taxes that would raise delivered IBA bauxite costs well above the costs of nonbauxitic ores.

In the short run, individual producing countries have substantial monopoly power for several reasons. First, alumina plants are, to a degree, specialized in their use of particular types of bauxite. Shifting to bauxite from a different source could, in some cases, impose large conversion costs. Individual countries could attempt to increase taxes or establish production minima to take advantage of this situation.

Second, the aluminum companies have substantial investments in the producing countries, and shifts to new sources of supply would require making further investments in the new supplying locations. Over periods of several years, the costs per ton of obtaining bauxite from a company's customary sources would be below the costs of bauxite from newly developed sources because of the additional investment required. Current host countries, such as Jamaica, might attempt to exploit this situation by raising taxes or production minima. Over periods of several years, depending on the economic life of the relevant facilities and the costs of constructing new facilities or adapting current alumina plants, some of the aluminum companies would have no real alternative but to pay the higher total taxes. The principal U.S. government policy option in such cases would probably be stock releases; such releases might require new statutory authority.

It is difficult to judge the likelihood of short-run monopoly pricing, whether unilateral or concerted. Although producing countries clearly have raised taxes and production minima, we believe it unlikely that they will exploit their short-run advantages so fully as to incur large risks of aluminum-company shifts to alternative sources of supply. The consequences for producing countries would be substantial and potentially disastrous.

Alternatively, the IBA might raise prices to a level well above the cost of nonbauxitic ores. The long-run effects of such an action on the United States would be equivalent to those of an embargo, and the long-run result might be a self-imposed U.S. embargo on IBA supplies. We believe such an action is highly unlikely, as it could cost the IBA its entire U.S. market. While stockpiles would provide a hedge against such an action, the major way to force down bauxite prices is to develop alternative sources of supply. Within four to five years, domestic bauxite and nonbauxitic sources and supplies from noncooperating countries could replace the high-cost IBA bauxite.

If the IBA believed that the combined actions of the U.S. government and the aluminum companies would cost it the U.S. market, short-run monopoly pricing would be unlikely. The long-run total cost to the IBA of losing the U.S. market is the present value of total taxes, which, of course, depends on the pricing scenarios. In any case, this cost runs into the billions of dollars and is vital to the economies of the poorer Caribbean countries. Since it is reasonably probable that the IBA will remain stable for a number of years, the Caribbean countries' future earnings may be relatively certain and not worth jeopardizing.

The IBA would, of course, gain from a short-run monopoly action; the potential gain would be no more than the resource loss the United States would suffer from a four-year 1980 embargo, or at most, $5 billion in present value as of 1980; in fact, the IBA would be unable to capture more than half of this cost to the United States. The IBA countries would suffer relatively great damages from losing the U.S. market, including the risk that, in the scramble to capture shares of the remaining world market, the cartel would collapse totally. Therefore, short-run monopoly pricing appears highly unlikely.

The likelihood of embargoes and extreme short-run price escalations is uncertain, and policy makers may wish to hold some level of stocks against these contingencies. While a strict economic calculus does not appear to justify holding economic contingency stocks much larger than present security stocks (equal to about nine months' consumption), much lower stocks could substantially raise the costs of a severe commodity action in bauxite. Therefore, economic contingency stocks on the order of at least six months' consumption, and perhaps as much as 12 months' consumption over and above those judged necessary for strategic purposes, may be justified. The risks do not justify large government programs to reduce import dependence or sustain new technologies, but do probably justify government research and development programs to reduce the costs of alternative supplies.

NOTE

1. See Charles River Associates, Policy Implications of Producer Country Supply Restrictions: The World Aluminum-Bauxite Market (Cambridge, Mass.: CRA, 1977), App. 5B, for detailed qualitative and econometric evidence on the demand for aluminum and substitution possibilities.

PART III

HIGH RISKS WITHOUT FORMAL CARTELS: CHROMITE AND MANGANESE

Risk, including both the likelihood and impact of interruptions, can be high even without a formal cartel. Major supplying countries might be politically friendly or unstable or unfriendly, and alternative foreign and domestically secure supplies might be unavailable. Furthermore, substitutes might be few and inefficient, and end uses might be critical to key industrial processes and employment. The chromite and manganese markets exhibit such tendencies to a significant degree.

The United States faces rather similar threats in both markets: the market structure and contingencies are broadly similar, and the flexibilities of consumption are small. Important differences remain, however, both between chromite and manganese and between these two commodities and others covered in this book.

Unlike the petroleum and bauxite markets, the chromite and manganese markets are not controlled by formal cartels. The major producing countries are so politically diverse that lasting explicit cartel formation is most unlikely. But both markets are highly concentrated. The Soviet Union and the Republic of South Africa both play major roles in them, raising the possibility of politically motivated supply interruptions. The importance of Rhodesia and South Africa makes the chromite situation a matter of particular concern. In both markets the major threats appear to be from possible short-term market withdrawals by one or more producers. Moreover, collusion among the remaining suppliers during a disruption by one major supplier could make the situation much worse for the United States. Limited tacit collusion among some major world producers is a continuing possibility, however.

Domestic land-based resources are relatively sparse and expensive for both, with the chromite-supply situation being particularly unfavorable. The most likely disruptions in these markets would be short-term, and they would be unlikely to provide sufficient incentives for development of domestic resources. The high cost and limited scope of domestic resources, particularly for chromite, mean that subsidies and many other supply-side policies are unattractive. Over a horizon which may extend beyond 1990, deep ocean nodules could provide alternative manganese supplies, reducing the long-term risk to the United States relative to that in the chromite market. The short-term risks remain very substantial in both markets. These risks are lowered by the large government and industry stocks of both materials—equivalent to over two years' consumption in each case.

It is sometimes argued that the current quantities of many materials, among them chromite and manganese, are required to prevent massive economic dislocations. If this were literally true, policy should be highly averse to risk and stocks very large. For both chromium and manganese we devoted considerable effort to determining possible substitutes and alternatives in end uses and in conservation opportunities. For both commodities we found modest but significant flexibility. The flexibility is, overall, considerably less than for copper or cobalt. Some manganese is currently considered essential to make high-quality steel as we know it; but the quantity used per ton of steel is not a technically immutable requirement, and there are substantial conservation opportunities as well as low-grade sources, such as the manganese which occurs naturally in many iron ores. Chromium is essential to the making of stainless steel. In some critical areas, such as oil refining, nuclear power, and jet engines, at present there is no reasonable substitute for stainless steel, and hence for chromite. However, this is far from true in all end uses, such as in largely decorative applications. Overall, there is at least some flexibility in stainless steel and chromite use.

For manganese, we investigated the techniques of steelmaking to identify technically feasible alternatives. For chromium, we integrated econometric estimates with the judgments of industry and other experts. Detailed investigation of the technologies involved also allows us to identify possibly fruitful areas for government involvement in technological-information consolidation and research.

Much of this effort was aimed at obtaining estimates of the price elasticities of demand for chromite and manganese—the estimated percentage by which consumption will decline, below what it would have been without the price increase, in response to a 1 percent increase in the material's price. Our estimate of the overall long-run price elasticity of demand for chromite is -.12 for substantial increases in chromite prices; that is, a 100 percent increase in price would cause only about a 12 percent decrease in use. Larger price increases are assumed to cause proportionately greater reductions in consumption, and in fact, the elasticity estimate -0.12 was selected to be most relevant to severalfold price increases. This means that users are willing to pay a great deal to avoid being forced to turn to less desirable alternatives. Small elasticities are indicative of relatively poor substitutes and of high possible disruption costs.

We based policy conclusions for the manganese and chromite markets on analysis of disruption scenarios, using optimal policy models of the type described in the Appendix. In both markets we first explored a range of pessimistic scenarios as potential worst cases. We also analyzed potential deterrence benefits from stocks and found that they could well be sufficiently important to increase substantially the efficient level of stocks. Our final estimates of economically efficient stockpiles, equal to 18-to-24 months of chromite consumption and 12-to-18 months of manganese consumption, were in fact based on scenarios exhibiting significant deterrence effects, both on the likelihood of embargoes and on the prices prevailing among them. Properly deployed, stocks substantially reduce the possible economic costs and dislocations of supply disruptions. However, we found that tariffs to decrease consumption prior to a disruption generally made only a small contribution in addition to that of stocks. Technological innovation to reduce consumption can yield substantial contingency benefits on the order of more than $10 million for each 1 percent decrease in consumption. Overall, we found that the possible impacts of supply disruptions justify real concern. But the United States is not without alternatives, and stockpiles, whether government or private, bolstered by some technological anticipation and preparation, could reduce the costs of even fairly severe crises.

6
DISRUPTIONS IN CHROMITE SUPPLIES: IMPACT AND POLICY ANALYSIS

Chromite, the ore from which chromium is obtained, was the first material designated as critical as World War II loomed. Chromium, used heavily in stainless steel, remains critical in a number of civilian and defense applications, including nuclear power, oil refining, and jet engine manufacturing. Forced decreases in consumption of chromite could have severe consequences for the United States.

This chapter presents impact and policy conclusions based on extensive research into the possible nature and size of such consequences and the alternatives available. Following general policy conclusions, we examine actual and potential domestic supplies, potential foreign supplies, and demand-side options; we emphasize throughout the potential policy implications. We then discuss the disruption scenarios and policy alternatives analyzed; particular attention is directed at stockpiling, tariffs, and technological adjustments. All of the analyses are based on the conclusions discussed in Chapter 3, namely, that a long-lasting formal cartel of chromite producers is much less likely than interruptions in supply from large suppliers as a result of either strategic or economic factors, possibly accompanied by informal or tacit collusion among remaining suppliers.

OVERALL IMPACT AND POLICY CONCLUSIONS

While the value of U.S. chromite imports is not enormous—around $200 million per year, including the value, as ore, of alloy imports—in some end uses, chromium is quite critical, and large shortfalls in supply would impose major dislocations or crises.

Unlike petroleum, however, chromite constitutes a small part of the cost of final products, amounting to less than 5 percent for stainless steel, the most critical and largest use. And, of course, stainless steel constitutes a very small share of the cost of most products bought by consumers. Even very large chromite-price escalations would not have substantial effects on typical product costs. Because the highly critical end uses constitute only a moderate proportion of consumption, supply shortfalls need not have disastrous consequences. Stainless steel accounts for over half of chromium consumption, but only about one-sixth of this use is in areas where at least moderately good substitutes are lacking. Substitutes for stainless steel in its less demanding end uses include plastics, coated and plated steel, titanium, and other materials. In the case of partial supply reductions, flexibility in chromium use in less critical applications could release enough supplies to sustain the more critical uses. Demand is relatively flexible in chemical uses such as pigments and plating, and substitutes are available in refractory applications. Overall, however, substitutions are sufficiently costly so that chromite demand is relatively insensitive to price; a doubling of price is estimated to cause only a 12 percent decline in consumption over a moderate time period.

We estimate that a 15 percent reduction in imports occurring with a 30 percent likelihood in any year (equivalent to a 97 percent likelihood of at least one disruption in a decade) could impose expected costs of about $1.5 billion, if no private or government stocks were available.* Such a severe reduction could result from

---

*Total expected costs stated in this chapter (and the chapter on manganese) are calculated at a 6 percent real rate of discount, in constant 1975 dollars, assuming the disruption threat remains the same indefinitely into the future. Expected costs are stated as a total present discounted value because such information is most relevant where the current cost of developing a technological adaption is to be compared with all future reductions in expected costs, which the adaption makes possible, in order to determine whether investment in developing the adaption is worthwhile. Alternatively, the <u>annual</u> expected cost, averaging years in which disruptions do and do not occur, can be obtained by multiplying the total expected cost by the real discount rate of 6 percent, in this case yielding (0.06) ($1,500 million) = $90 million per year, equal to approximately 45 percent of the annual value as ore of U.S. manganese imports and consumption.

a withdrawal of Rhodesian supplies, coupled with constraints on South African exports. A more severe, 26 percent disruption, as might result from the disruption of both Rhodesian and South African supplies occurring in 30 percent probability, imposes expected costs of as much as $3.2 billion when there are no stocks. Of this, over $1.2 billion is the cost to users forced to shift to less desirable alternatives.

Efficient stockpiles substantially cut the disruption costs to users. The actual gains may be even higher, as wasteful accelerated investments are reduced, and unemployment and other consequences are avoided. As discussed below, contingency stockpiles equaling a range of 18 to 24 months' consumption can be justified under pessimistic assumptions about potential disruptions.

Optimal tariffs are on the order of 15 percent of normal competitive prices. These tariffs make little contribution to reducing disruption costs, in part because stockpiling is such a flexible policy, but basically because U.S. supply and demand are so insensitive to price that tariffs (and quotas) reduce imports very little. Tariffs cannot be recommended strongly: they will be resisted by users, and their potential contribution can be realized almost fully by the more flexible stockpile tool with its greater potential for deterrence effects. Unlike the situations in the oil and copper markets, domestic chromite production costs are so high that the tariff would induce little or no production and create no vested interest, except perhaps among tariff-dependent producers of substitutes. Because of these high domestic production costs and small domestic resources, we did not consider subsidies a feasible policy alternative, and did not explicitly evaluate them.

We evaluated the contingency benefits of technical adaptations that would decrease imports. For the 15 percent disruption discussed above, saving 10,000 tons of chromium consumption annually (about 2 percent of total consumption) yields a present value contingency benefits of about $22 million. For the 26 percent disruption, the saving is about $39 million. If stocks were not available, the benefits of technologies reducing consumption would be even larger. Examples of projects reducing consumption include the development of surface implantation of chromium to replace stainless steel and the recovery of chromium chemicals from tanning and electroplating solutions. Increased secondary recovery of stainless steel could have similar benefits. Such conservation projects might be implemented at moderate costs and could yield benefits even in the absence of a supply disruption. The government might coordinate information gathering relative to such subjects.

Overall, significant stockholding appears to be the best policy measure. Massive programs to develop domestic production would be wasteful: the resource base does not exist. Imports might possibly be reduced by the solution of various technical problems, including surface implantation of chromium and the welding of stainless steel sandwich materials.

Various noneconomic policies have clear potential. The U.S. government could encourage the exploration and development of new chromite sources. In particular, Brazilian expansion might be encouraged. Monitoring of the political situation in southern Africa surely is continuous, and U.S. policy toward emerging regimes in that region might consider problems of chromite supplies, not to mention cobalt from Zaire, and platinum and manganese from South Africa.

CHROMITE SUPPLY: DOMESTICALLY SECURE AND POTENTIAL FOREIGN SOURCES

We will consider domestic supplies in decreasing order of the speed they offer in responding to possible disruptions: stockpiles, secondary-chromium recovery, and primary resources.

Domestic Stockpiles

U.S. industry and government together hold very large stocks of chromite, amounting to over four years' supply. Stocks throughout the critical metallurgical industry were, in the 1975-76 period, about six months' consumption, and private stocks of refractory and chemical-grade materials were significant. Government stocks are very large; prior to 1976, GSA held, in excess of strategic objectives, almost three years' consumption of chromium in metallurgically useful forms. However, strategic stockpile goals were raised sharply in October 1976 to the point that a deficit approximately equal to a half year of U.S. primary consumption was created. Also, not all private stocks are available for contingency use because some are required as working inventories. The private sector, particularly metallurgical users, evidently believes that shortfalls in chromite supply are sufficiently likely and damaging to justify substantially costly precautions, such as the holding of large stocks.

Domestic Secondary Recovery

Recycling already contributes to U.S. chromium supply. Most secondary chromium is recovered from stainless steel scrap; it amounts to approximately 10 percent of primary chromium consumption, but it appears that this supply would be very unresponsive to chromite-price increases. In a very severe supply crisis, the supply of scrap could be increased, though none of the disruptions we considered formally appear sufficiently severe to make such wartime efforts worthwhile. If sustained high prices for chromium were likely, programs to facilitate recycling probably would be worthwhile. For example, stainless steel parts could be stamped with their alloy composition and designed for easier removal when junked.

Increased secondary recovery from materials other than stainless steel is also a possibility. One source has estimated that about 11,000 tons of chromium in chemicals are discarded annually in tanning solutions, and that it would not be a difficult research task to discover how to recover this material if chromite prices became high enough.

Domestic Primary Chromite Resources

Domestic chromite resources are quite sparse and do not represent a substantial long-run supply alternative. Indeed, the chromite resources of the United States are not only quite small, but of low grade compared to most ores that reach the world market. Although chromite was mined in the United States until 1961, since 1900 annual production was significant only in years when it was heavily subsidized by the government. Even during emergency war years, domestic production satisfied only a small percentage of demand. Chromite mining in the United States ceased in 1961 when the government stockpile-purchasing program was terminated. Scattered information on the domestic costs of producing chromium suggests that a price of approximately $800 per short ton of contained chromium (in 1975 dollars) might be sufficient incentive for domestic production to reach the peak rates achieved during the subsidized government-stockpiling program of the 1950s—about 40,000 short tons of contained chromium per year, or about 7 percent of normal U.S. consumption.

Although chromite prices may be temporarily high enough to justify exploitation of poor domestic resources during a severe foreign supply disruption, it is highly unlikely that prices would stay high enough for long enough to justify the substantial fixed investments required. The capital for mining, processing, and transporting chromite is very durable and requires long lead times to bring onstream. There would be a great risk for investors that the chromite price would drop long before the benefits of these investments justified their acquisition costs.

If a highly stable cartel were formed from all the countries with rich chromite resources, especially South Africa, the Soviet Union, and Rhodesia, lower-grade deposits in other countries would put an upper bound on the price that could be charged by the cartel in the long run. The long-run upper bound provided by resources of the quality of those in the United States would apparently be at least twice the late-1975 price for Soviet ore, which was itself roughly double the reasonable estimates of a long-run competitive world price.

Our review of reserves data shows that deposits in foreign noncartel countries would be more economical to develop than those in the United States. Also, chromium resources in other countries typically have been much less intensively explored than those in the United States, so increased prospecting due to higher prices is more likely to uncover exploitable deposits elsewhere. Overall, domestic resources represent a poor prospect, both as a current alternative and as the basis for policies such as production subsidies.

Supplies from Nonrestricting Foreign Countries

A formal and complete chromite cartel is most improbable: South Africa, the Soviet Union, Southern Rhodesia, Turkey, Albania, India, Finland, and Brazil are not all likely to cooperate. Both econometric and geologic evidence indicates that at a sufficiently higher price, supplies in many areas can be expanded substantially, though not always with great speed. In the event of a severe disruption by one or more major suppliers, there would be great incentives to expand production in other countries. However, a localized crisis would also offer an improved opportunity for tacit collusion and price escalation among remaining suppliers.

CHROMITE DEMAND: POTENTIAL FLEXIBILITY IN USE AND TECHNOLOGICAL ADJUSTMENTS

Except for stocks, domestic chromite supply options are unfavorable and uneconomic. As a result, demand-side adjustments assume particular importance. If there were no flexibility in chromite use (that is, if all chromium used were essential), then it might pay to hold extremely large stocks. Our investigations indicate that neither extreme is strictly true. Some uses are indeed critical; the function served is essential, and substitutes are poor, so that the costs of making adjustments are high. Our basic aims were to identify both the more and the less flexible end uses, and to quantify roughly the consumption adjustments that would be made in response to various degrees of escalation of chromite prices. To do this we integrated a wide range of econometric and engineering evidence to obtain a realistic appraisal.

A major message from this work is that it can be misleading to talk in terms of overall chromite and chromium use. We found that in general, chromite use is quite insensitive to price changes, but end uses do differ significantly in flexibility. Thus, under the impact of a supply crisis and high prices, the more flexible uses, such as many in chemicals, would be able to turn to substitutes, freeing scarce chromium supplies for the less flexible and more critical end uses.

Below, we first briefly discuss the technical factors underlying the various uses of chromite. Then we consider the relative importance of the various end uses, along with summary information on the price sensitivity of demand. Finally, we discuss the technical considerations underlying these results.

Technical Basis of Chromium Use

There are several reasons why chromium alloys, particularly stainless steel, are used extensively in industrial and consumer products. The principal reason is chromium's high corrosion resistance, which is generally unmatched by any other element at comparable costs. Chromium's corrosion resistance is maintained at very low and high temperatures. This combination of heat and corrosion resistance makes chromium products wear well in a wide variety of environments, including rain, sun, chemicals, water, and high temperatures. Chromium also provides an attractive surface to items which are coated with it, and stainless steel derives its well-known shiny quality from chromium. As this range of characteristics suggests, the potential substitutes range widely, depending on the desired product attributes.

Relative Importance of the Major End-Use Categories and Demand Elasticities of Chromium

It is usual to distinguish three major end-use categories for chromium: metallurgical, chemical, and refractory. In 1973, total metallurgical uses accounted for 76.6 percent of chromium consumed in the United States, about three-fourths of which was used in stainless steel production; chemical uses accounted for about 11.8 percent, and refractory uses accounted for 11.6 percent. The price elasticities of chromite demand differ significantly among the end uses.

Table 6.1 presents our basic elasticity estimates, along with the relevant market shares. The critical fact to note is that stainless steel, the largest single use category, is also the least elastic. For moderate price increases, considerably more chromium would be released from uses other than stainless steel, despite its accounting for the majority of chromium consumption.

TABLE 6.1

Chromite-Market Shares and Long-Run Price Elasticities of Demand

End Use	Elasticity of Demand	Market Share (percent)
Stainless steel	-0.04	57.4
Other metallurgical	-0.12	19.2
Refractories	-0.15	11.6
Chemicals	-0.46	11.8

Source: Authors' estimates, discussed in Charles River Associates, Policy Implications of Producer Country Supply Restrictions: The World Chromite Market (Cambridge, Mass.: CRA, 1976).

Metallurgical Uses of Chromite: Substitution Possibilities and Technological Alternatives

Virtually all of the chromite for metallurgical uses is first converted to one of a variety of chromium-intensive alloys. These alloys are then used in making stainless steel and other alloys, which are themselves used in various final and intermediate products. We will examine the substitution possibilities at each stage, giving special attention to stainless steel.

Production of Ferrochromium

The main issue in the production of ferrochromium is the ability to use relatively low-grade, high-iron (chemical) ores from South Africa in place of higher-grade (metallurgical) ores from Rhodesia, the Soviet Union, and Turkey. In the United States ferrochromium generally has been produced from the metallurgical-grade ores. Technical changes, which have made use of the lower grades easier, include the Japanese solid-state reduction of chromium (SRC) process, which also results in lower loss of chromium. Increased flexibility in choice of ore grade would lower the impact of disruptions of supplies from any one country, particularly Rhodesia and the Soviet Union.

Industry sources have indicated that many U.S. ferrochromium producers could switch to lower-grade South African ores without major investments. These shifts could become profitable if supplies of Soviet, Rhodesian, or Turkish metallurgical grades were seriously disrupted or became much more expensive. A problem is that a less valuable high-carbon, high-iron product results. However, the widespread introduction of the Argon-Oxygen Decarburization (AOD) process in stainless steel production over the last decade has increased the demand for these lower-grade ferroalloys. Recent developments reported by Union Carbide suggest that the more desirable low-carbon product can be produced from the lower-grade ores. Little is publicly known about these techniques, but the result is increased supply flexibility. And, over the long run, substantially higher chromite prices could induce development of techniques to reduce chromium losses during ferrochromium production.

The Demand for Ferrochromium in Steel Production

Most ferrochromium is used in stainless steel. Other uses include alloy steel, cast iron, and superalloys; of these other uses, all but superalloys use quite small amounts of chromium per unit of output.

Massive adoption of the AOD process has allowed greater flexibility in chromium use in stainless steel production; with this process, a much larger proportion of high-carbon ferrochromium can be used, and hence the potential costs of disruptions in Soviet and Rhosedian supplies are reduced. The process has improved efficiency in chromium recovery, as only 5 percent of the chromium in the ferroalloy is lost, though during a severe crisis some of this loss might be reduced slightly. A newer but basically similar method is the Creusot-Loire-Uddenholm (CLU) process.

If chromite became much more expensive, it might become economical for producers to cut back on ferrochromium use by more carefully controlling production practices, adding closer to the minimal amount needed to achieve desired product characteristics. Other technical changes might achieve the same effect: our analysis of manganese use, for example, disclosed that relatively inexpensive production changes could result in significant manganese conservation.[1]

We now turn to possible ways of cutting down on chromium (ferrochromium) consumption for the production of stainless and other steels and alloys.

Chromium Substitution and Conservation in Stainless Steel Production. Stainless steel as a use category accounts for some of the most critical and least flexible major uses of chromite, and hence is the area of greatest potential crisis impact. There are limited ways to reduce the intensity of chromium use in the making of stainless steel, and, more importantly, substitutes exist for stainless steel in some applications.

There are bounds on the amount of other materials that can be substituted for chromium and still produce a steel with physical properties comparable to those of a given grade of stainless steel. In general, chromium is the most effective single additive to steel for the purpose of imparting resistance to corrosion, particularly at high temperatures. No other alloying agents have been able to match the properties which it adds to steel, at anything near a competitive price. Corrosion resistance in steels falls rapidly when chromium content is reduced below certain levels, notably, about 14.5 percent. In most uses requiring significant corrosion resistance, a typical 500-grade steel (5 percent chromium) could not efficiently substitute for a typical 18 percent chromium grade.

Several materials, including nickel, molybdenum, aluminum, silicon, and cobalt, impart some corrosion resistance to steel, and so are possible chromium substitutes. Nickel, cobalt, and molybdenum are all at least four times as expensive per unit of contained element as high-carbon ferrochromium, and together or singly they do not come close to imparting the same degree of corrosion resistance.

Most of the decrease in chromium use in stainless steel would occur on the demand side as some users switched to stainless steel substitutes or otherwise decreased their use of stainless steel. We now consider this issue.

Reducing Chromite Consumption by Using Substitutes for Stainless Steel. Drastically higher chromite prices or supply problems will raise the cost and price of stainless steel, or, at least equivalently, make supplies of it unavailable and uncertain. In either case, some users will wish to shift to substitutes. These shifts for some applications will reduce the total demand for stainless steel and hence for chromite, freeing supplies for uses with higher economic priorities. We estimated the possible adjustments in stainless steel use by grouping consumption according to the severity of the working environment. We then obtained, based on discussions with industry experts, consensus-judgmental estimates of possible adjustments to varying magnitudes of stainless steel price escalation within each category. These estimates were aggregated to obtain an overall estimate of the price elasticity of demand for stainless steel.[2]

Therefore, to estimate the aggregate price elasticity of demand, we divided the end uses of stainless steel into three categories according to the severity of environment in which the stainless steel is used. The most crucial reason for choosing stainless steel over other materials is its great resistance to corrosion at elevated temperatures. As the severity of the environment increases, the types of substitutes available and the feasibility of using them change. In general, the more severe the environment, the less varied and potentially efficient the possible substitutes. Examples of uses for stainless steels that were categorized by severity of environment are listed below.

Construction and Contractors' Products

- gutters and downspouts
- swimming-pool accessories
- storm windows and doors
- fixtures
- playground equipment
- wash basins
- drinking fountains
- handrails
- louvers
- mullions

General-Purpose Industrial Products

- boiler equipment
- heat exchangers
- steam separators
- pumps
- valves
- clutches
- diesel and gas engines
- mechanical power-transmission equipment
- lubricating machinery
- ball and roller bearings
- pulleys

Electrical Machinery and Equipment

- turbines
- motors and generators
- telephone equipment
- transformers
- transmission hardware

Other Domestic and Commercial Equipment

- furniture
- cabinets
- curtain rods
- vending machines
- laboratory apparatus
- toys
- home incinerators
- filing cabinets
- typewriters
- surgical equipment
- precision tools
- photographic equipment

Table 6.2 allocates the market shares of these end-use groupings among three different categories, based upon the severity of the application's environment: mild, harsh, and very severe.

Category I (mild environments) includes products which generally are used under relatively mild conditions, such as contact with water, food, or sunlight. Stainless steel is chosen for these applications because it is resistant to contamination by bacteria, it does not fade or rust quickly, and/or it is shiny and pleasant in appearance. We estimate that about half of stainless steel consumption falls into this category, including decorative automotive uses, construction, appliances, and food-processing equipment.

TABLE 6.2

U.S. Stainless Steel Consumption by Environmental-Severity Category
(percentage)

Category I (Mild)		Category II (Harsh)		Category III (Very Severe)	
Type of Use and % in Category	% of Total 1973 U.S. Consumption	Type of Use and % in Category	% of Total 1973 U.S. Consumption	Type of Use and % in Category	% of Total 1973 U.S. Consumption
Automotive (75%)	10.7	Automotive (25%)	3.6	Aircraft (100%)	2.9
Construction and contractors products (100%)	13.1	Other domestic and commercial equipment (50%)	0.6	General industrial equipment (50%)	3.1
Other domestic and commercial equipment (95%)	11.9	General industrial equipment (50%)	3.1	Electrical machinery equipment (50%)	1.9
Appliances, utensils, and cutlery (100%)	9.9	Bolts, nuts, rivets, and screws (100%)	3.4	Other special industrial (50%)	1.4
Food processing equipment (100%)	2.7	Electrical equipment and machinery (50%)	1.9	Metal-working equipment (50%)	0.8
Total	48.3	Metal-working equipment (50%)	0.8	Chemical-industry equipment (25%)	0.6
		Chemical-industry equipment (75%)	1.8	All other (50%)	7.1
		Textile equipment (100%)	1.2	Total	17.8
		Pulp and paper equipment (100%)	1.2		
		Shipbuilding and marine (100%)	0.9		
		Industrial fasteners (100%)	4.2		
		Other special industrial equipment (50%)	1.4		
		Forgings, not elsewhere classified (100%)	2.5		
		All other* (50%)	7.1		
		Total	33.7		

*Includes equipment for nuclear plants, sewage treatment, pollution control, mining, agriculture, construction and building, oil and gas drillings, and hand tools.

Note: Categories are as defined in the text. The total percentages add to 0.998 because of rounding error.

Source: Authors' estimates, derived from data in International Nickel Company (INCO), "Stainless Steel Market Studies," 1967 outlook and summary; Interim Review of 1967 Forecast and New Forecast 1970; Individual Industry Outlooks and Five Year Summary (1974).

Category II (harsh environments) includes uses in which resistance to significantly corrosive environments is required. These uses include continual contact with acids, other chemicals, sea water, gases, petroleum, and abrasives. Included are the catalytic-converter portion of automotive uses and a wide variety of industrial equipment. About one-third of stainless steel use falls into this category.

Category III (very severe environments) includes contact with corrosive agents at extreme temperatures and under severe stress, proximity to nuclear reactions, and contact with molten metals. About one-sixth of stainless use falls in this category.

Our general finding is that close to half of stainless use is in relatively mild environments where its unique properties are not really essential. Less than 20 percent of stainless steel use is in very severe environments; these uses might be considered critical.

Table 6.3 shows estimated demand responses to stainless steel price increases of 10 percent, 25 percent, 50 percent, and 100 percent. For the purpose of investigating the impact of realistic chromite-price changes on the stainless steel market, only the first and, possibly, the second rows in the table are relevant. Since chromite typically accounts for less than 4 percent of the price of stainless steel, the price of chromite would have to increase by a factor of more than 3.5 to cause a cost-justified or based increase of 10 percent in the price of stainless steel.

The percentage increases in the price of stainless steel should be considered relative to the prices of potential substitutes. Thus, the price changes indicated in the first column of Table 6.3 correspond to percentage changes in the current dollar price of stainless steel only if the prices of potential substitutes stay approximately constant. The potential substitutes and the basis for our demand estimates are discussed below.

Substitutes for Stainless Steel: Category I (Mild Environments)

There are a number of substitutes for stainless steel in the applications grouped into Category I, including anodized aluminum, carbon steel coated with epoxy or glass, and various types of plastic such as lexan, polyesters, and polyethylene. Stainless steel is used in many of the products in Category I because it presents an attractive appearance and does not rust. The competitive materials may not be as attractive and/or as corrosion resistant as stainless steel, but they are generally priced considerably lower. One manufacturer of food-processing equipment reported that his firm had recently switched to stainless steel from anodized aluminum and

TABLE 6.3

Estimated Long-Run Demand for Stainless Steel at Higher Prices by Category, Relative to Mid-1975 Levels

Amount of Stainless Steel Price Increase above 1975 Level Relative to Prices of Substitutes[a]	Proportion of Base-Period Demand Remaining after Long-Run Adjustment to Price Increase		
	Category I (weight: .483)[b]	Category II (weight: .337)[b]	Category III (weight: .178)[b]
10%	0.9	0.95	0.99
25	0.75	0.875	0.975
50	0.5	0.75	0.95
100	0.3	0.6	0.80

[a]1975 average price is $.66 per pound.

[b]These weights are the proportion of total base-period demand accounted for by each of the stainless demand categories, as derived in Table 6.2.

Note: See Charles River Associates, Policy Implications of Producer Country Supply Restrictions: The World Chromite Market (Cambridge, Mass.: CRA, 1976), App. 4B, for an explanation of how the estimates in this table are combined into a single price-elasticity estimate.

Source: Charles River Associates, Policy Implications of Producer Country Supply Restrictions: The World Chromite Market (Cambridge, Mass.: CRA, 1976), p. 134.

epoxy-coated steel. A 10 percent increase in the price of stainless steel would not cause the firm to shift back, but a 25 percent increas probably would, and higher increases would certainly induce a substitution.

A materials expert for a U.S. automobile manufacturer stated that there are viable substitutes available for almost all automotive applications of stainless steel, including the catalytic converter. Aluminized steel and chromium-coated steel are usable in catalytic converters without substantial loss in performance, and aluminum or coated plastics can be used for wheel covers, hubcaps, bumpers, and decorative trim. A 10 percent increase in the price of stainless steel could potentially cause a significant shift away from stainless steel, while higher price increases would certainly cause such a shift.

In applications such as appliances and other domestic and commercial equipment, stainless steel is often used solely on the basis of its appearance, as only minimal corrosion resistance is required. A 10 percent increase in the price of stainless steel could be expected to cause some substitution, although a larger price increase would be required to induce massive substitution.

Based on such information, a 10 percent price increase in the price of stainless steel is estimated in Table 6.3 to cause a 10 percent decrease in the long-run demand for stainless steel in Category I. Price increases of 25 and 50 percent are also estimated to cause equal percentage decreases in demand.

Substitutes for Stainless Steel: Category II (Harsh Environments)

In the range of relevant prices, fewer substitutes that can compete efficiently with stainless steel are available in Category II than in Category I. While some of the same materials mentioned for Category I are also considered substitutes in some Category II applications, the loss in performance and durability is considerably greater than for the Category I uses. For example, one chemical company indicated that it would continue to specify stainless steel for its tank trucks even if the price rose by 25 percent, because the life expectancy of a stainless tanker is 12 years, whereas that of an uncoated carbon steel tanker is 18 months. Glass-coated steel would be a more viable substitute, but the glass eventually peels away and must be replaced at a high cost.

Substitute materials that would come closer to matching the performance of stainless steel in some Category II uses include copper-base alloys such as cupronickel; nickel-base alloys such as monel; certain alloy steels, and titanium. All of these materials generally have been more than twice as expensive as stainless steel, except for the alloy steels, which also contain considerable amounts of chromium. Since stainless steel provides adequate corrosion resistance for uses considered here, other materials typically would not capture the bulk of stainless steel's market share, even in the event of a relative price increase as large as 50 percent.

Another possibility for reducing the use of stainless steel in this category is the technique known as cladding regular carbon steel with stainless steel. This method can reduce the amount of stainless steel required by as much as 90 percent, for example in large pressure vessels where a thick core of plain carbon steel substitutes for pure stainless. Stainless clad steels are not widely used currently because the technology for elimintating leaks at the weld lines is not yet perfected. Fabricating costs are higher than for pure stainless steel, but the total cost of stainless clad steels can be significantly lower where thick steel is required, because most of the stainless is replaced by cheaper carbon steel. Because cost saving from using stainless clad steels declines as the thickness of the required steel declines, there are many applications where clad steels would never be efficient even with anticipated technological advances. A major increase in the price of stainless steel would generate much interest in clad steels; according to one industry source, the technological improvements needed to make clad steels reliable and efficient in many more applications are currently under investigation and are expected to be developed in the near future.

A 10 percent stainless steel price increase is estimated in Table 6.3 to cause only a 5 percent decrease in the Category II demand for stainless steel—approximately half the price elasticity estimated for Category I. Again, proportionately larger decreases in the quantity demanded are specified for price increases of 25 and 50 percent, though it is not anticipated that even severe crises in the chromite market would cause such large price increases for stainless steel, or if so, only for brief periods.

Substitutes for Stainless Steel: Category II (Severe Environments)

For many of the applications of stainless steel in Category III, the only substitutes available are at least twice as expensive as stainless steel, and sometimes are three to four times as expensive. Some of these materials, however, are, in certain respects, superior to stainless steel.

Titanium competes with stainless steel in applications which involve severely corrosive environments and very high temperatures as well as severe stress. Examples of these uses include seawater desalinization plants, jet engine parts, offshore oil wells, and nuclear-power plants. Depending upon the application, pure titanium costs from three to eight times as much per pound as stainless steel, but since structurally equivalent titanium weighs considerably less than stainless steel, the cost differential is actually somewhat less. Titanium has already replaced stainless steel in some extremely severe environments, but in most cases, stainless is probably in more direct competition with some of the nickel-base superalloys, which are also closer to its price range. Titanium is not really a potential bulk substitute for stainless steel in the short run, because of limited manufacturing capacity. In the long run, reserves of oil from which titanium metal can be produced (rutile and, at slightly higher cost, ilmenite) are plentiful and production of titanium metal could be greatly expanded at approximately constant cost.

Several industry sources confirmed that for some of the applications in Category III, there are no substitutes for stainless steel because of its unique combination of properties. One such application is the catalytic cracking unit in petroleum refineries, for which corrosion and heat resistance, in addition to significant strength, are needed. Stainless is not as strong as carbon steel, but is significantly stronger than titanium and some of the other nonferrous alloys.

For stainless steel used in applications in Category III, a 10 percent price increase is estimated in Table 6.3 to cause only a 1 percent decrease in the quantity demanded. Decreases in stainless steel usage in this category contribute only negligibly to the price elasticity of demand for chromite over the range of higher prices considered in the supply-disruption scenarios.

Based on a cost-pass-through calculation, the stainless-demand estimates of Table 6.3, and minor adjustments to allow for possible shifts to lower chromium-content stainless grades and reduced use of chromium per pound of stainless, we estimate the long-run price elasticity of demand for chromite in stainless steel to be about -0.04. That is, we estimate that a 100 percent increase in the chromite price would reduce total use in stainless steel by only 4 percent. This small response is due in part to the small share of chromite in total stainless cost, and in part to the high cost of using substitutes or replacements. Much larger chromite-price increases or recurring supply problems could cause more dramatic shifts. As we have seen, substitutes, though costly, exist in most applications.

Chromium Flexibility and Conservation in Other Alloy Uses

Other alloy uses of chromium are diverse and account for about one-third as much chromium consumption as does stainless steel, namely, about 19 percent of total consumption. Because of this diversity, and because any individually critical uses account for a very small proportion of the total, our investigation was considerably less detailed for other alloy uses than for stainless steel.

The principal use categories are alloy steels, tool steels, and superalloys. Chromium is added to alloy steels to make them harder—not corrosion resistant—and in most cases, reasonably efficient substitutes are available, including boron, nickel, and molybdenum. Because the chromium content of these steels is low, about .25 to 1.5 percent, even massive increases in chromite prices would have only minor price effects on the price of finished steel.

Tool steels contain relatively more chromium, generally 3 to 5 percent, which is added for hardness, heat resistance, and some degree of corrosion resistance. While vanadium grades of steel could substitute for chromium alloys, alloying substitutes are generally not closely competitive in price. However, ceramics and carbides compete strongly in many applications.

Considerable amounts of chromium are used in superalloys, for many of the same reasons that chromium is used in stainless steels. While no good chromium substitutes are available in superalloy use, surface implantation of chrome could help in some applications. Chromium is added to some cast iron to impart hardness. The proportions of chromium in cast iron are very small, and, because corrosion resistance is unnecessary, nickel, molybdenum, and tellurium are effective substitutes. Finally, a small amount of chromium goes into nonferrous alloys, where materials such as tungsten can substitute. Since the relatively flexible alloy-steel use accounts for about half of total use, we judged that chromite use in these alloys was more flexible than in stainless steel, and assumed a price elasticity of demand of -0.12.

Flexibility in Chemical Uses of Chromite

Some of the chemical uses of chromite are also based on chromium's corrosion resistance. In total, they account for approximately 12 percent of total consumption. The main use categories are plating, pigments, and tanning. More than half of plating applications are decorative, such as auto bumpers; plastics, nickel, cadmium, zinc, or ceramics can substitute in decorative uses. About one-quarter is in hard plating for abrasion and corrosion resistance; the only good substitutes in these uses are other chrome-bearing products, such as stainless and alloy steels. The final quarter of plating uses is in anodizing and metal finishing, where various substitutes exist.

Chrome pigments are used for their stable nature, under corrosive and hot conditions, such as in highway and industrial uses. Potential, although inferior, substitutes include cadmium yellow. Because industrial processes do not strongly depend on these pigments, they cannot be deemed critical uses.

Most common leather items are tanned with chromium solutions, but demand has declined as leather has been replaced by its substitutes. Vegetable tanning is an alternative, although it is less effective than chrome tanning, and at much higher chromite prices, zirconium salts are an alternative.

The above uses account for about 70 percent of chemical applications. The remainder of the chemical consumption is divided among numerous small end uses, most of which have good potential substitutes. We use an econometrically estimated long-run price elasticity of demand of -0.46 for chromite in chemical uses. This elasticity is much greater than in the other categories, consistent with generally more favorable substitution possibilities.

Flexibility of Chromite Use in Refractories

About 12 percent of chromite consumption is used to produce refractory bricks to line high-temperature furnaces, mainly open-hearth and electric furnaces. Refractory use has decreased because of the adoption of basic oxygen furnaces, which do not use chrome refractories. Refractories can be made from a variety of ore grades, though the Philippine high-aluminum type is typically the best for this use. Magnesite bricks, or bricks containing a higher proportion of magnesite to chromite, are an actual substitute, and in fact are superior for higher, more variable temperatures. Magnesite is, however, more costly than chromite at normal prices. Substitutes are available, and we conservatively assume a long-run elasticity of -0.15 for chromite demand in this use. The actual flexibility is probably greater than indicated by this estimate.

The Overall Flexibility of Chromite Demand

Weighting the elasticity estimates for the end uses, we obtain an estimate of about -.12 as the price elasticity of chromite demand for moderate price increases over the intermediate long run. That is, we estimated a doubling of chromite price would cause a 12 percent decrease in consumption below what it would otherwise have been. The estimate is conservative, as it is based to a large extent on currently known technology. Sustained price increases for the very long run could allow more economizing, but the estimate is reasonable for likely disruption scenarios. Uncertainty about the duration of disruptions could inhibit some long-lived investments in such areas as retooling, and so it is proper to be conservative in estimating possible substitutions.

SUPPLY-DISRUPTION SCENARIOS AND POLICY OPTIONS

We estimated optimal economic contingency-stockpile levels for situations with and without deterrence effects, whereby increased U.S. stocks decrease the probability of disruptions occurring or make U.S. imports less costly during disruptions. Without deterrence effects, efficient sizes of stockpiles are in the range of one year's consumption or less, even when disruptions are considered to be likely.

However, contingency stockpiles of as much as two years' consumption can be justified assuming what appear to be moderate deterrence effects, as further discussed below. The Appendix contains a further discussion of deterrence effects and illustrates stockpile calculations for the chromite market.

Supply Disruptions without Deterrence Effects

Our initial policy evaluations are based on two supply disruption scenarios. The less severe of the two involves a 15 percent decrease in U.S. chromite and chromium-ferroalloy imports; the more severe involves a 26 percent disruption. While South Africa and Rhodesia supply between 40 and 50 percent of the non-Communist-countries' consumption, even elimination of these supplies need not reduce imports by 15 percent, since other supplying countries would be induced to expand supplies. The two disruption scenarios are, however, realistic possibilities. They could result from varying degrees of political disruption in southern Africa, combined with tacit price collusion by the Soviet Union and Turkey. Since the likelihood of a disruption is uncertain, we analyzed the two scenarios for several sets of disruption probabilities. In the base case, Scenario A, the probability that a disruption will occur in any given year (the "inception probability") is 0.3, implying a 97 percent probability that a disruption will occur at least once during a decade. The probability that a disruption will continue for another year (the "continuation probability") in the base case is 0.5, and hence the probability that a disruption, once begun, will last five years or longer is 0.16. Realistically, if a disruption of this magnitude lasted five years or longer, the expansion of alternative supplies would probably reduce its severity to well below 15 percent by the end of the period.

We also analyzed a more optimistic case, Scenario B (0.2 inception and 0.4 continuation probabilities), and a more pessimistic case, Scenario C (0.4 inception and 0.6 continuation probabilities). All of the scenarios are very pessimistic about the chances of a disruption; even in the more optimistic Scenario B, the probability that at least one disruption will occur over a decade is 90 percent. By focusing on such pessimistic assumptions, we are, in effect, building a great deal of risk aversion into our impact and policy analysis.

Welfare Costs without Deterrence Effects

As an upper bound, we estimated expected costs of disruptions, assuming no policy preparation or stockpile releases. For the 15 percent base-case disruption, the expected cost is about $1.5 billion, of which about 78 percent is increased payments to foreign producers and 22 percent is economic-disruption costs to users. These are expected present value constant-dollar costs for all future disruptions, expressed in 1976 dollars and discounted to 1976 at an annual real discount rate of 6 percent, and weighted by the probabilities of disruptions. These calculations of expected costs assume that the threat of a disruption, both in terms of its magnitude and its probability, remains the same indefinitely into the future. For the 26 percent disruption, the total expected cost is $3.2 billion, of which about 38 percent is disruption costs to users forced to resort to high-cost substitutes. In the 15 percent case, the total expected cost ranges from slightly over $1 billion in the more optimistic scenario to about $1.9 billion in the more pessimistic case. For the 26 percent disruption, costs to users are a substantially greater proportion of total costs, because without stocks, significant amounts of the supply shortfall will come out of the more critical uses, for which substitutes are less effective.

During disruptions of these magnitudes, market prices could rise dramatically, particularly if stocks were unavailable. The formal policy model predicts that without stocks, prices would rise more than eightfold in the first year of a 15 percent disruption and more than threefold in the very long run in such a disruption. For a sustained 26 percent disruption, without stocks, the policy model predicts almost a seventeenfold increase in the first year and over a fivefold increase in the very long run. These predictions are probably an overstatement of what would happen during disruptions of these magnitudes, to the extent that inventories were available though speculative price escalations could be greater; on the other hand, the estimates may be biased downward because the demand curves of the policy model are designed to give accurate estimates of the cost of conserving chromium at the lower prices that would occur if sizable stock releases could be made. Nevertheless, these calculations are a useful way of putting into perspective the implications of the very low price elasticity of U.S. demand for chromium, both in the short run and long run.

These estimates are very pessimistic in that they assume that neither government nor private stocks are available. While policy decisions would be needed to make present government stocks available to industry, private stocks will of course be used if it appears profitable to do so. By assuming that no stocks are available, this analysis considers users to be at the mercy of higher prices and so overstates the possible impacts of disruptions. In fact, users do hold substantial stocks.

Economically Efficient Levels of Contingency Stocks, without Deterrence Effects

We applied the optimal policy program—assuming stockpiles have no deterrence effects—to the scenarios described above. Efficient stock levels are those which minimize expected future disruption costs, allowing for the costs of holding stocks. Without deterrence effects, these efficient stocks are relatively modest, even assuming that the probability of a disruption occurring is high.

For the 15 percent supply cutback, without deterrence effects, we found efficient stocks to equal about six months' normal consumption in the base case, and to range from about 4.2 to 6.8 months' consumption for Scenarios B and C. For the 26 percent disruption, efficient stocks were much larger, ranging from over nine months' worth of supplies to about 14 months' worth for Scenarios B and C, and totaling almost a year's consumption in the base case.

Releasing these stocks reduces the likely cost of possible disruptions. For example, the base-case economic-dislocation costs for a 15 percent disruption are over $320 million when there are no stocks, but fall only about $53 million when the efficient stock level equal to six months' worth of supplies is deployed. This reduction in expected costs is by a factor of more than 6. In the 26 percent disruption scenario, economic-disruption costs to users are over $1.2 billion when there are no stocks, but fall to only $108 million when the efficient stock level, equal to about a year's normal consumption is available. Stockpiling thus has a high potential benefit-cost ratio. More important than the reduction in our estimated expected costs, access to stocks allows users to avoid making many costly adjustments that would be reversed after a disruption ended. In addition, the psychological impact of the embargo is lessened, and possible political effects are much reduced.

These efficient stockpile levels are of course dependent on the scenarios assumed. In general, our scenarios tend to assume relatively likely (and possibly recurring) but short-lived disruptions. Inception probabilities of 0.3 per year mean that the probability of <u>no</u> disruption over a decade is below 0.03. Under these assumptions, disruptions are virtually certain to occur. If the probability of a disruption's continuing another year is 0.5, then the probability of a disruption's lasting more than five years is only about 0.06. Political changes, such as anti-Western regimes in Rhodesia or South Africa, might cause longer-lasting disruptions. We evaluated the effects of such possibilities on our estimated results by determining efficient stockpiles for scenarios with low probabilities of occurrence but high likelihoods of continuation. For a 26 percent disruption with a 0.02 yearly inception probability and 0.95 continuation probability (equivalent to an 18 percent probability of a disruption's occurring within a decade, and an 81 percent probability of a disruption's lasting over five years once begun), the efficient level of stocks is about ten months' worth of supplies.

Supply-Disruption Scenarios with Deterrence Effects

Disruptions caused by political turmoil in southern Africa or by shifts in Soviet export policy are not likely to be deterred initially by larger U.S. chromite stockpiles. However, a disruption in southern Africa is quite likely to encourage collusion among remaining suppliers, notably, the Soviet Union and Turkey, and larger U.S. stocks might well discourage this collusion, reduce prices, and increase supplies. In addition, for most disruption scenarios, it is plausible to assume that larger U.S. stock releases during disruptions will result in a lower price on the world market and hence lower prices for the remaining amount of U.S. imports. Thus, the larger efficient stockpiles that are calculated under assumptions of probability-deterrence effects and price-deterrence effects may well be more reliable guides to policy making than the levels obtained ignoring such effects. Note, however, that our assumptions about the strengths of these effects are only illustrative, and are not based on explicit economic (or political) analysis.

We first analyzed the possible importance of deterrence effects in the context of the 15 percent disruption with the 0.6 continuation probability, as discussed above. If there were no deterrence effects, the efficient level of stocks for a 0.05 yearly disruption inception probability would equal about three and one-half months' normal consumption. The expected disruption cost is about $417 million. If approximately doubling stocks to about seven months' supplies reduces the yearly embargo probability to 0.03, the expected welfare cost (with the costs of the extra stock included) is $300 million, for a net deterrence gain of $267 million. Such effects could yield handsome benefit-cost ratios for stockpiling policy. Of course, smaller deterrence effects imply smaller benefits from increasing stockpile sizes, or possibly even losses. If doubling stocks in the same basic scenario reduced the yearly embargo probability from 0.05 to only 0.048, the added stocks would not cover their costs.

The Appendix to this volume discusses the generalized optimal policy model we used to calculate efficient stockpile sizes where deterrence effects exist. Making what appear to be reasonable assumptions about deterrence effects on disruption likelihoods and prices, it is possible to justify a chromium contingency stockpile equaling between 18 and 24 months' consumption, assuming an annual inception probability of roughly 0.04 per year (equivalent to a 0.34 probability of a major disruption over the course of a decade). The calculations of optimal chromite stockpiles above and in the Appendix are intended to be representative of the analyses which lead us to our final conclusions regarding desirable stockpile sizes in the chromite market. However, these illustrative calculations are not a complete justification in themselves; indeed, it would be desirable to do much further analysis of this type before final policy decisions are made.

U.S. Chromium Stocks and Possible Supply Disruptions

In mid-1975 U.S. producer and consumer stocks of chromium equaled about eight months' typical consumption. Government stocks in excess of strategic objectives contributed another 40 months' consumption, for a total of over four years' worth of supplies. Effective available supplies were somewhat less, mainly because of needs for working inventories, but clearly, stocks were large. Total stocks in fact appeared larger than would be efficient under most assessments of contingency risks in the absence of very strong probability-deterrence effects. However, in October 1976 strategic stockpile objectives were raised to the point where a deficit approximately equal to a half year of U.S. consumption was created. Thus, the amount of chromium stocks in the United States ostensibly available for nonmilitary emergencies became less than we would recommend.

Tariffs as an Adjunct to Stockpiling Policy

Tariffs on chromium imports would reduce imports prior to an embargo, thereby lowering consumption and encouraging other preparations for reduced supplies. We considered tariffs in conjunction with stockpiles and determined the levels of both policy instruments that, taken together, would yield the largest saving in embargo costs, allowing for both the costs of holding stocks and the costs of tariff-induced distortions. Tariffs are, in general, not a very efficient policy, because demand is not highly sensitive to likely price increases; stocks provide a more efficient cushion for critical uses. In part, the tariff benefits are small because stockpiling is such a useful, flexible policy; overall, tariffs on chromite are not a desirable or highly beneficial option.

We found that efficient tariffs to reduce consumption were, in most cases, about 15 percent of estimated normal prices of chromite. Under the most favorable assumptions, tariffs yield small reductions in expected disruption costs, only on the order of several million dollars. These gross benefits may not even be sufficient to cover the administrative costs of the tariff, implying that the policy would not be worthwhile.

Benefits of Technological Preparations for Supply Disruptions

Reducing Consumption

A number of technological changes could reduce chromite consumption and possible crisis impacts. These changes include perfecting welding methods for stainless steel clad materials, developing surface implantation of chrome, and increasing secondary recovery. However, chromium users would object to incurring higher costs for long periods of time by introducing technological changes that were not commercially efficient at normal prices.

In the 15 percent disruption scenarios, a development reducing consumption by 1 percent reduces the welfare cost (with no stockpile or tariff) by between $10.3 and $19.4 million, depending on the likelihood of the disruption and assuming no deterrence effects. For the 26 percent disruption scenarios, the loss reductions are between about $22.1 and $41.4 million. Stockpiles can reduce the benefits somewhat, to about $8.4 to $15.9 million in the 15 percent case, and $14.5 to $27.4 million in the 26 percent case. Note that these are gross contingency benefits that do not take account of the costs of making the adjustments and do not include normal economic benefits during nondisruption periods.

We have not evaluated specific projects, but our analysis of demand and substitution possibilities indicated a number of promising areas, as mentioned above. Whether pursuing these projects would be desirable will depend on actual costs and crisis likelihoods. Contingency benefits as calculated here are but one part of the overall calculation of costs and benefits. Government activity might include, for example, gathering information about and perfecting fabrication methods for stainless clad materials.

Increasing the Speed of Adjustment of Consumption to Long-Run Possibilities

Our disruption-cost estimates assume a relatively slow adjustment of demand to the long run; full adjustment is assumed not to occur until the eighth year of a disruption. This period of adjustment is reasonable since many of the substitutes will be costly and difficult to implement. Increasing the speed of adjustment will lower disruption costs. For base-case probabilities, with available stocks, a 10 percent increase in adjustment speed yields contingency benefits of about $0.76 million; for the 26 percent disruption, the corresponding benefits are about $1.6 million. These increases in adjustment speed would reduce slightly the size of the economically efficient contingency stockpile.

The main conclusion from our contingency-benefit estimates is that such benefits are probably small relative to the likely magnitude of the expense of increasing the speed of adjustment in U.S. demand. Increasing the speed-of-adjustment factors by 30 percent would be a massive project, almost certainly costing more than the largest estimated contingency benefit of $4 million. It is hard to conceive of an individual project being justified solely by contingency benefits derived from this type of effect, although such benefits might be a significant addition to a project which derives benefits from other effects.

It is difficult to give specific examples of what might be required to increase the speed of adjustment of domestic demand to higher chromium prices. Information about alternative nonchromium materials to which consumers could switch after a chromium-supply disruption developed seems to be widely known in the industry. Programs to prepare the required retooling and increased capacity in industries which supply chromium substitutes could be undertaken, prior to the occurrence of any disruption, and pilot projects to demonstrate the feasibility of such switches could be funded by the government. However, the political implications of the government's helping aluminum, plastics, and titanium producers to take over stainless steel markets, even on a contingency basis, seem rather complex. Moreover, the contingency-benefit estimates suggest that, in most cases, a more efficient policy would be simply to ensure more postdisruption adjustment time by means of a larger stockpile, and to begin converting to substitutes only after a serious disruption in chromium supply develops.

NOTES

1. See Chapter 7 of this book, and Charles River Associates, <u>Policy Implications of Producer Country Supply Restrictions: The World Manganese Market</u> (Cambridge, Mass.: CRA, December 1976), pp. 72-116, 125-32. Because of the large number of apparently nonessential uses of chromium, we did not analyze these issues in detail but focused on substitutes. To the extent that such economizing possibilities are important in current uses of chromium, our results understate the flexibility of chromite consumption and overstate the possible effects of supply restrictions and the efficient policy preparations, such as stockpiling. Since the worst danger in such studies would be to understate the possible impact, the possible bias is made on the conservative side. Chromite is generally a more significant cost element than manganese, so substantial economizing may already have occurred. This subject clearly merits some investigation.

2. We had several reasons for pursuing this approach rather than the standard one of statistical demand estimation. The data on stainless steel use and prices are not rich enough in price variation to allow isolation of price effects from other factors, such as technical change. In addition, data are simply too sparse for extensive and reliable application of such techniques to research. Also, as discussed in Charles River Associates, <u>Policy Implications of Producer Country Supply Restrictions: A Framework for Analyzing Risks, Imports and U.S. Government Policies</u> (Cambridge, Mass.: CRA, 1976), pp. 72-75, estimated functions will reflect historical price-quantity relations, while our concern is with the unusually large possible price escalations that might occur during a supply crisis. Perhaps more important, exploring the technology of use can indicate areas for research or other government action.

7
MANGANESE: IMPACT AND POLICY ANALYSIS

For all practical purposes, manganese from high-grade ores is an essential input into the modern steelmaking process, which in fact uses over 90 percent of the higher-grade manganese ores that are mined. Ubiquity of use in all basic steelmaking distinguishes manganese from chromite, whose use is concentrated in relatively fewer areas. Disruptions of U.S. manganese supplies would directly affect more sectors of the economy than is the case with chromite, but the increase in the cost of final goods and services is likely to be relatively less in individual cases. Although disruptions causing severalfold increases in the price of manganese ore are serious possibilities worthy of careful attention from policy makers, realistic scenarios do not entail a complete cutoff of manganese that would bring U.S. industry to a standstill.

As discussed in Chapter 3, there are a somewhat greater number of major manganese suppliers than is the case with chromite. The most important supplier, South Africa, accounts for about one-third of non-Communist supplies, with Australia, Gabon, and Brazil accounting for significant percentages. In the event of a disruption in South Africa and/or Gabon, some expansion in supplies from Australia may be possible. U.S. stocks are relatively large, so there is no necessity for use to decline suddenly during an emergency of moderate duration, though government strategic stocks may not be readily available. Also, the amount of manganese required per ton of steel is not rigidly fixed by technology; our investigation of the technology of steelmaking shows that the four- or fivefold manganese ore-price increases that could accompany a serious supply disruption could induce economies of over 10 percent in the short run and more than 35 percent in the longer run. Finally, deep-sea manganese nodules provide an essentially unlimited supply source in the very long run.

Because manganese ore contributes a very small share to the cost of most steels, about 0.3 to 0.4 percent, even large increases in manganese prices would have, at most, small effects on steel prices and use. Consumers probably would not perceive the effects on prices of final goods.

Our policy analysis found that the significant costs of shortfalls justify holding an economic stockpile equaling roughly 12 to 18 months of consumption. Modest government programs to ensure wide access to information on manganese conservation could ease adjustments, as could some basic research in these areas. Government actions such as a partial price-support program could help accelerate and assure the development of ocean mining. However, resolution of outstanding international disagreement over the ownership of ocean resources is more important than either price supports or programs to encourage information distribution and research.

In the remainder of this chapter, we first evaluate potentially secure supplies of manganese. Domestic resources are not promising, and ocean nodules are not a current or short-term option. We also evaluate demand-side options, including replacements for manganese. The analysis identifies a number of manganese-conserving adjustments in steelmaking technology, and quantifies the potential extent of conservation in response to price escalations. Using this information, we then estimate the welfare impacts of a range of supply-disruption scenarios and evaluate stockpiling, tariff, and technological policy options.

SECURE SOURCES OF MANGANESE SUPPLY

We first consider domestically secure sources of manganese, in rough order of the speed of response they would offer in the event of a disruption: stocks, recycling, land-based deposits, and ocean mining. We then briefly consider possible expansion of some sources of foreign supply.*

---

*Manganese is imported into the United States in a number of forms, notably, high-grade ores and the ferromanganese or other refined products used directly in the steelmaking process. Unless otherwise specified, the discussion of manganese in this chapter refers to high-grade ores or equivalents. Foreign producers basically control ore sources, though they are also increasingly refining ore to ferromanganese before export. Many iron ores contain

Domestic Stocks of Manganese

Domestic stocks of manganese are quite large. Not including government stocks held to meet mid-1976 strategic objectives, industry and government stocks totaled over 3.3 million tons of contained manganese, or over 27 months of recent consumption. However, in October 1976 strategic goals were raised to the extent that government excess stocks dropped to the equivalent of only about a half year of U.S. consumption; the government excess and private stocks together then totaled only about 15 months of U.S. consumption.

Industry stocks themselves typically account for close to ten months of consumption. While some industry stocks will always be needed as working inventories, private stocks alone would provide a substantial cushion in case of a supply emergency. Full use of the government stocks would involve some extra costs, as some are of relatively low grade or are stored in distant locations. Overall, though, domestic stocks could provide a quick response to offset even massive import disruptions, although the release of government strategic stocks would be required to ameliorate the impacts of protracted disruptions.

Recycling and Secondary Recovery

Large amounts of manganese are lost in the steelmaking process. The total amount of manganese present in steel produced in the United States is probably less than the total amount of manganese originally present in various steelmaking inputs other than the ferroalloys produced from high-grade manganese ores. The most important of these sources of manganese is iron ore. The manganese from iron ore and scrap iron is by and large lost, predominantly into the slags of the blast and steel furnaces.

Manganese can be recovered from slags in two ways. First, slags could be processed into synthetic manganese ores. This proved uneconomic in the 1950s and there have been no technological breakthroughs since. Such production could only prove economic at much higher manganese prices.

Second, the slag can be recycled in the production process.[1] Slag recycling generally involves adding steel-furnace slags to

---

substantial quantities of manganese; most of this manganese is normally lost into the slags of the steelmaking process. There are some ores traded that, along with iron, contain less than 35 percent manganese and are therefore not considered high-grade manganese ores.

blast-furnace contents; about 75 to 80 percent of manganese in steel-furnace charges goes into the slag. Currently there is limited use of slag recycling in the United States. At normal prices, the value of the manganese savings is overshadowed by the savings of other recoverable materials and by the reduced needs for lime and other materials, including fluorspar. These benefits are significant, and it appears that some slag recycling may be economical at current price levels, although recycling does impose increased capital and fuel costs. The extent to which slags can be recycled is constrained by the resulting buildup of phosphorus and chromium in the steel. However, the manganese savings are significant, and higher manganese prices could accelerate the recycling of slags.

Domestic Land-Based Manganese Resources

The United States has no manganese reserves, that is, no deposits known and efficiently mineable under current technological and economic conditions. Domestic manganese resources, as opposed to reserves, are large and estimated to contain over 40 years' current consumption. These resources are quite low in manganese content and would be very expensive to extract.

Spurred by above-market-price purchases and other government aids, including loans and special purchases from small producers, domestic production of manganese peaked in 1957 and declined thereafter, ceasing after 1970. Even at the 1957 peak, production was well below 20 percent of consumption. The last U.S. manganese mine capable of producing any significant output at commercially competitive prices had largely ceased operations by 1960. Production at this mine was subsidized by approximately 75 percent over the 1959 market price, which appeared to be near long-run competitive levels, and production by various small producers was subsidized by approximately 130 percent over the 1959 market price.

Today, much larger subsidies would be required to yield substantial production. The U.S. mines formerly producing at commercial prices apparently have been largely depleted, so moderate subsidies would not result in their reopening. The quality of known, undeveloped U.S. resources has almost certainly not increased as much as the quality of exploited foreign reserves has increased since 1959. The decline in the world real price of manganese has been partially due to technological advances in large-scale mining that may not be completely transferable to the lower-grade U.S. deposits. Increases in real mining costs in the last several years have been disproportionately high due to higher energy costs, and recovery of manganese from lower-grade ores, such as those found in the United States, is relatively more energy intensive.

Overall, domestic land-based resources appear to be a poor base for much production. During a disruption, prices might briefly rise high enough for such production to yield contingency benefits. However, reintroduction and expansion of production requires a long planning horizon, and the resulting capital is very durable. As a result private investment in domestic manganese resources during a disruption would appear unprofitable and hence be unlikely; there also appears to be no strong rationale for the government to undertake such investment.

Production from Deep Ocean Nodules

The U.S. Geological Survey estimates potential resources from deep-sea nodules at over 5.2 billion tons in gross weight. At an average manganese content of 20 percent, the nodules contain over one billion tons of manganese, which equals more than a century of world consumption at current rates. Depending on the prices of other metals contained in the nodules, principally nickel, copper, and cobalt, extraction of manganese could be profitable at prices somewhat above current levels. Most trade sources consider nickel, cobalt, and copper the only economically recoverable resources in the nodules, though at least one consortium has assumed that manganese recovery will contribute to revenues.

The ownership status of the nodules is still uncertain and depends on the resolution of a number of international political questions. This factor and remaining technological uncertainties suggest that undersea sources of manganese should be regarded as speculative for at least the next decade. Potential production could provide a significant check on sustained increases in the price of land-based supplies.

Supply Expansion by Relatively Secure Foreign Sources

The nature of the major Australian deposit allows sustained large-scale production and somewhat more rapid expansion than is possible elsewhere. It appears that Australian government policies will allow continued development, and that Australia would be unlikely to restrict supplies sharply.

Resource depletion makes Brazil's production of higher-grade ores uncertain. However, new deposits reportedly have been discovered within the last two years, further exploration is under way, and research is being conducted to use abundant low-grade resources. Brazil has not been eager to restrict exports of other materials, although expansion of Brazil's domestic steel industry could limit manganese exports.

DEMAND-SIDE ALTERNATIVES: REPLACEMENTS FOR MANGANESE AND ECONOMIES IN MANGANESE USE

Except for stocks, there are few short-run domestic supply alternatives for manganese. Demand-side options therefore assume particular importance. We concentrate here on the use of manganese in iron and steel production; other end uses account for less than 10 percent of consumption. Some manganese is used in nonferrous alloys with copper, aluminum, and other metals, and nonmetallurgical uses include dry cell batteries and chemicals used in fertilizers, paint, varnish, and glass production. Manganese compounds can be used as antiknock additives and may eventually be used as catalysts to purify automobile exhausts. Uses other than steelmaking are likely to remain relatively small. Following a brief discussion of the function of manganese in steelmaking, we address possible replacements for manganese in these uses. We then turn to ways of economizing on manganese use in steelmaking, summarizing our results in terms of the overall price flexibility (elasticity) of demand for manganese. While the elasticity estimated is small, manganese use is not totally rigid.

The Functions of Manganese in Steelmaking

Manganese is added to steel in large part because of its effects on steel cleanliness and hot working properties as well as on steel heat treating and hardenability. Liquid steel contains dissolved oxygen and sulfur, which impart undesirable properties to the steel. Manganese performs the essential roles of taking sulfur and oxygen into the slag, or of combining with the sulfur and oxygen in more benign forms in the final products. In particular, manganese combines with sulfur so that steel is not made overly brittle. At manganese prices well above normal levels, sulfur control would be the dominant consideration in determining minimum acceptable manganese additions.

While manganese also improves the mechanical properties of steel, it is not added to ordinary steels for this purpose. The manganese content of most steels is under 1 percent. Certain special manganese alloys contain up to 14 percent manganese. In some applications, such as auto bumpers, aluminum-deoxidized steel can replace the high-manganese steels. Some manganese is also used in stainless steel production. However, overall, common carbon steel accounts for the bulk of manganese consumption.

Several steelmaking trends could increase the use of high-grade manganese ores per ton of steel. Shifts to iron ores of lower manganese content, such as taconite, could increase the need for manganese additions. Use of higher-sulfur cokes could have the same effect. Both of these factors contribute to the high Soviet usage of manganese, which is perhaps twice the non-Communist countries' average use per ton of steel. Higher costs for fluorspar, a flux agent widely used in steelmaking to help remove sulfur and phosphorus, have led to research into methods to substitute manganese for fluorspar. If these trends continue, manganese use per ton of steel could increase significantly, although higher manganese prices or shortages could reverse this trend.

Material Replacements for Manganese in Steelmaking

Although material substitutes for manganese in steel exist, not only are they much more expensive and potentially limited in supply, but they are also more troublesome to use than manganese. Such replacement materials as calcium and magnesium are difficult to add to molten steel, and they are generally very volatile, so that little of them is actually retained in the steel. Titanium and zirconium affect heat-treating possibilities and so are limited in application. Some substitutes, particularly rare earths, have advantages in special steel types, although these materials are expensive and have properties not needed in most applications.

Manganese is overwhelmingly cost effective in the bulk of steelmaking. It is most unlikely that supplies of material replacements could be expanded rapidly, and these expanded supplies would be available only at costs much higher than the current price of manganese.

Methods for Economizing on Manganese Use

Technical adjustments in processing that could conserve on manganese use include external desulfurization, continuous casting, production controls and related adjustments in steel specifications, and adjustments in the steel product mix.

External Desulfurization

In external desulfurization, sulfur is removed from molten iron after it is produced in blast furnaces and before it is charged into steel furnaces. External desulfurization cannot eliminate completely the need for manganese additions. Nevertheless, external desulfurization could allow manganese use in steel to be reduced by as much as 25 percent; to achieve this reduction, the steel-specification ranges for tolerable maximum sulfur and minimum manganese contents would have to be adjusted. This would not generally lead to a lowering of steel quality. The government might play a role in assuring rapid adjustment of steel specifications to reflect this change.

Our calculations indicate that the ferromanganese savings from external desulfurization would be worth about $1.00 per ton of steel, while external desulfurization costs about $3.40 per ton of steel (1976 prices). A doubling of manganese prices would provide a noticeable, though not overwhelming, increase in incentives for external desulfurization. Another possible incentive is the fuel-cost saving allowed by increased use of higher-sulfur coals. However, EPA regulations currently limit use of such coal and so may slow adoption of the technique.

Continuous Casting of Steel

In continuous casting, steel is cast into continuous lengths, not discrete ingots. Less manganese is required in this process for adequately hot working properties than in conventional methods. The process now accounts for about 10 percent of production, but may account for 40 percent by 1985. Continuous casting yields other benefits, including a shorter production process, savings in space, and the absence of ingot molds and other equipment. But the massive capital requirements involve lengthy planning and construction. Higher manganese costs would be only a small incentive to accelerate the introduction of continuous casting, and the lag in the response would be very substantial.

Production Controls and Changes in Steel Specifications

Manganese historically has constituted a very small fraction of total steel costs. Producers have consequently had little incentive to economize on consumption, and many steel producers have used relatively inexact manganese-addition procedures. The procedures used are designed to insure consistency with the manganese specifications of the American Iron and Steel Institute (AISI) which typically allow the manganese content of steel to vary within a range such as 0.30 percent to 0.60 percent. Producers have typically aimed at the midpoint of the allowable range.

Steel products can fail to meet manganese specifications in two ways. First, the molten product may have had too much or too little contained manganese; that is, the ladle analysis (manganese content) of the molten metal after ferroalloy additions may be too high or too low. Second, manganese content may vary from one part of the ingot to another, as metals tend to segregate during cooling. If a sample taken from part of an ingot has an unacceptable manganese content, the ingot will be rejected. This variation in product-analysis is affected by cooling conditions, such as choice of ingot mold. Continuous casting allows less segregation of manganese during the cooling of steel; hence, manganese content could be reduced with less risk of not meeting specifications for the minimum allowable amount of manganese.

The degree of ladle-analysis variation is significantly determined by the packaging and handling procedures for ferroalloys. Because rejection of a batch of steel is very costly, and because product analysis also varies, producers are generally conservative, and add sufficient manganese ferroalloys to aim for the middle of the allowable specification range for the cooled steel. Production controls that could reduce variation in the ladle analysis include alloy packaging, whereby manganese alloys are placed in sized boxes or measured with automatic machinery, reducing error; alloy sizing, in which alloys are crushed and screened to size prior to boxing, thus reducing uncertainty about amounts added; and agglomeration or canning of fines, because finely powdered materials tend to disperse in the furnace.

These changes could become efficient if ferromanganese prices doubled relative to the price of other steelmaking inputs, and they could be adopted relatively quickly. Much of the equipment for sizing/packaging and automatic handling is in place in newer facilities.

Manganese-specifications ranges for steel are not directly determined by the required physical properties of the product. To a significant extent, they are determined by institutionalization of the range of product analysis, which has been typical for the industry. After several years, changes in ladle-addition practices and other changes allowing reduction in manganese usage, such as external desulfurization of hot metal, might, in the normal course of events, result in new specifications.

Changing steel specifications is not a costless procedure, partly because of the required reeducation of consumers. Special government programs for adjusting steel specifications quickly were instituted during World War II, and these same programs could be used during a peacetime emergency, including dissemination of information to steel producers and consumers. It would presumably not be absolutely necessary to plan such a program of industry consultation and communication before the occurrence of an emergency. However, accumulation of basic information on technological procedures allowing production of quality steels with less primary-manganese consumption should be an ongoing process. Where benefits of these projects would accrue to industry as a whole, government support might well be justified.

Adjustments in the Steel Product Mix

The level of manganese consumption is affected by changes in the mix of steel produced, since stainless and alloy steels contain considerably more manganese than the carbon steels. Substantial increases in manganese costs could affect this mix, although on average, the cost of ore required for manganese additions (usually as ferromanganese) represents only about 0.3 percent of the value of steel. A tripling of manganese-ore costs would raise average steel costs by just 0.6 percent. In some cases, changes in the relative prices of steels due to increased manganese costs could affect the choice of steel types, stimulate the development of new alloys, and reduce the intensity of manganese use to below what it otherwise would have been. However, since the bulk of manganese is used in the production of plain carbon steel, the effect of differential steel-price increases on the overall mix of manganese contents in the short and intermediate run is likely to be minimal.

The Effects of Higher Prices on Manganese Use in Steelmaking

We synthesized the various technical adjustments outlined above into estimates of possible manganese conservation induced by higher prices. The estimates are a distillation of expert opinions about the nature of a complete manganese-conservation program. As such, they cannot be definitive or precise. Rather, they are a first approximation to the information required for effective policy making. The analysis uses as a bench mark a doubling of the ferromanganese price, equivalent to multiplying the ore price by about 4.7, other things equal.

Within a year after such dramatic increases in the prices of manganese ore and ferromanganese, better production controls and other adjustments could allow total consumption to decrease by about 10 percent. Among other adjustments, ferromanganese producers could economize on manganese, some steel users could shift among grades, and some users could shift away from steel. Direct use of higher-grade manganese ore in pig iron production would probably diminish rapidly.

If high manganese prices persisted for a number of years, the above adjustments would probably become industry practice. In addition, some AISI manganese-specifications ranges could be reduced. Over the long run, investment in external desulfurization and continuous casting, combined with possible further adjustments in some steel specifications, could, we estimate, reduce manganese use to about 65 percent of its original level relative to the amount of steel production. Uncertainty about the duration of higher prices could prevent producers from making manganese-conserving adjustments, just as it could deter investment in domestic production. Investment in external desulfurization is a substantial undertaking, and investment in continuous casting is an even more massive one. Such investment would not be undertaken in response to short-run incentives, although it should be noted that these processes are expected eventually to become standard industry practice in any case. We have assumed that the manganese savings from such adjustments in the capital stock are realized slowly, over the horizon of eight years. Conservation measures involving production controls require only minor capital investments and hence can be instituted over a shorter time horizon.

SUPPLY-DISRUPTION SCENARIOS

We have analyzed two major sets of supply-disruption scenarios for the manganese market. Some cases include deterrence effects, whereby increased U.S. stocks decrease the probability of disruptions or make U.S. imports less costly during disruptions.

Contingency stockpiles equaling more than a year's consumption can be justified assuming what appears to be plausible deterrence effects, as discussed further below.

Analysis of Supply-Disruption Scenarios without Deterrence Effects

As in the chromite market, we first analyzed 15 and 26 percent import reductions under various combinations of yearly crisis-inception and -continuation probabilities, assuming stockpiles have no deterrence effects. Let us first consider a disruption due to political events in South Africa. In 1974 South Africa accounted for 20 percent of world production and supplied over 30 percent of non-Communist consumption. However, even a complete elimination of South African production might not reduce supplies reaching non-Communist consumers by over 15 percent for more than a year or two, if remaining suppliers increased their production in response to the sharp price increases resulting from the disruption.

The import reduction could be maintained at 15 percent, or even somewhat more, if other countries took advantage of the tight market for economic or other gain. For example, India and Gabon might impose export taxes, and Brazil might impose an export quota ostensibly to protect dwindling reserves. However, a significant part of the South African supply cutoff could be offset by Australian expansion. The disruption could average 15 percent over roughly five years if, quite implausibly, South Africa were entirely cut off for that long. A 26 percent disruption of U.S. manganese imports was considered in addition to the 15 percent disruption.

The 15 percent disruption was analyzed using three different sets of disruption probabilities, ranging from likely (Case A), to more likely (Case B), to most likely (Case C). The annual inception and continuation probabilities for these three cases are summarized in Table 7.1. Even in less pessimistic Scenario A, the probability of no disruption over a decade is only 35 percent. However, the scenarios do assume that most disruptions are relatively short; even in most pessimistic Scenario C there is only a 13 percent probability that a disruption will last five years or more.

Expected Welfare Losses in the Absence of Government Policies

Assuming that no stocks, public or private, are available, the expected economic cost of a 15 percent disruption scenario ranges from about \$560 million for Scenario A to \$1.5 billion for the most pessimistic case, Scenario C. These are present value constant-dollar expected costs for all future disruptions, expressed in 1976 dollars, discounted to 1976 at an annual rate of 6 percent, and weighting disruption costs by probabilities. In the base case, Scenario B, the total expected costs are about \$1.1 billion, of which about 20 percent is adjustment costs to users. The threat of disruptions, both in terms of probabilities and severity, is assumed to be maintained indefinitely into the future in making these calculations. In the 26 percent disruption, the expected costs range from about \$1.0 billion for Scenario A to \$2.8 billion for Scenario C. In Scenario B the costs are \$1.9 billion, of which 35 percent is user adjustment cost. These cost estimates are biased upward since they were calculated assuming that neither private nor government stocks were made available. On the other hand, we have not taken full account of the likely trauma and dislocation of planning that such disruptions could easily cause. If no stocks were available, some steel plants would be forced to cease operations, at least temporarily, and workers would lose their jobs.

TABLE 7.1

Manganese Supply-Disruption Probabilities

Scenario	Probability That a Disruption Will Begin in Any Year (yearly inception probability)	Probability That a Disruption Will Continue Another Year (yearly continuation probability)	Probability of Disruptions within a Decade	Probability That Disruption Will Last 5 Years or More, Once Started
A—disruption likely	0.1	0.4	.65	0.03
B—disruption more likely (base case)	0.2	0.5	.89	0.06
C—disruption most likely	0.3	0.6	.97	0.13

Source: Statement of assumptions used for scenarios considered in Chapter 5 of Charles River Associates, Policy Implications of Producer Country Supply Restrictions: The World Chromite Market (Cambridge, Mass.: CRA, 1976).

During disruptions of these magnitudes, market prices could rise dramatically if stocks were unavailable. Our formal policy model (described in the Appendix) predicts that without stocks, prices would rise six- or sevenfold in the first year of a 15 percent disruption and two- or threefold in the very long run. For a sustained 26 percent disruption without stocks, the policy model predicts a ten- or elevenfold increase in the first year and a three- or fourfold increase in the very long run. These predictions are an overstatement of what would happen during disruptions of these magnitudes, to the extent that inventories were available; on the other hand, the estimates may be biased downward because the demand curves of the policy model are designed to give accurate estimates of the cost of conserving manganese at the lower prices than would occur if sizable stock releases could be made. Nevertheless, these calculations are a useful way of putting into perspective the implications of the very low price elasticity of U.S. demand for manganese, both in the short run and long run.

Efficient Levels of Contingency Stocks without Deterrence Effects

We first applied our optimal policy program to the scenarios described above, assuming stockpiles have no deterrence effects. Efficient stock levels are those which minimize expected future disruption costs, allowing for the costs of holding stocks. Without deterrence effects, these efficient economic stocks represent less than one year's consumption, even if the probability of disruptions occurring is high.

For the 15 percent supply disruption, the most efficient stock levels ranged from 2.5 months' worth of consumption (Scenario A) to 4.8 months' worth (Scenario C). In the base scenario (Scenario B), the optimal level of stocks equals almost four months' worth of consumption. It reduced disruption costs to the United States from more than $220 million to about $50 million. For the 26 percent disruptions, efficient stocks ranged from six months' worth of consumption with Scenario A probabilities to over nine months' worth with Scenario C probabilities. For Scenario B probabilities, efficient stock levels of about eight months' consumption reduced expected U.S. costs from $668 million to $108 million. All of these scenarios are pessimistic, because they assume that disruptions are highly likely to occur. For comparison, we found that in an alternative 26 percent disruption scenario involving less likely, longer-lasting disruptions (0.01 inception, 0.9 continuation probabilities), efficient stocks were a bit under three months' worth of consumption.

Supply-Disruption Scenarios with Deterrence Effects

Disruptions caused by political turmoil in southern Africa or elsewhere are not likely to be deterred initially by larger U.S. manganese stockpiles. However, subsequent collusion among remaining manganese suppliers might well be deterred by larger U.S. stocks. For most disruption scenarios, it is plausible to assume that larger U.S. stock releases during disruptions will result in a lower price on the world market and hence lower prices for the remaining amount of U.S. imports. Thus, the larger efficient stockpile sizes calculated assuming probability-deterrence effects and price-deterrence effects are very probably a more reliable guide to policy making than the estimates discussed above.

We first analyzed the possible importance of deterrence effects in the context of the 26 percent disruption, with a 0.9 continuation probability—the very pessimistic scenario mentioned above. If there were no deterrence effects, the efficient level of stocks for a 0.05 yearly inception probability would be equal to about 16 months' normal consumption. The total expected disruption cost is about $2.1 billion. If raising stocks by 60 percent, to about 25 months' supplies, reduces the yearly embargo probability to 0.03, then the expected welfare cost (with the costs of the extra stocks already netted out) is $1.6 billion, for a net deterrence gain of $0.5 billion. Such large deterrence gains could obviously justify the expense of holding a much larger stockpile.

Making reasonable assumptions about probability-deterrence effects and price-deterrence effects, it is possible to justify a manganese contingency stockpile equaling between 12 and 18 months' consumption, using annual inception probabilities of between 0.03 and 0.04. (An inception probability of 0.035 per year is equivalent to a 0.3 probability of a major disruption occurring over the course of a decade.) Price deterrence effects of stockpiles are likely to be relevant for most disruption scenarios, and they can be predicted with reasonable accuracy from examination of models of the manganese market. Price deterrence effects alone, with little or no probability deterrence, can justify stock sizes equal to a year of consumption. Moderate additional probability deterrence effects can easily raise the optimal stock size to 18 months of consumption. These issues are discussed further in the Appendix, using examples from the chromite market.

U.S. Manganese Stocks and Possible Supply Disruptions

In 1977 U.S. producer and consumer stocks of manganese equaled about ten months' typical consumption. Government stocks in excess of strategic objectives contributed another six months' worth for a total of 16 months of supplies. Effective available supplies may be somewhat less, mainly because of needs for working inventories. If excess government strategic stocks were made available during a supply disruption, then total available stocks would be near the low end of the range of efficient contingency stocks discussed above. However, there is no guarantee that excess government strategic stocks would in fact be made available in a nonmilitary emergency.

The Merits of Efficient Tariffs in the Disruption Scenarios

As in the chromite market, we found tariffs for manganese to be relatively ineffective; demand responses to moderately increased costs are not large, and tariff-induced distortions are costly. As in chromite, efficient tariffs would not induce significant domestic production of manganese. In the scenarios considered, the efficient tariff is generally about 10 percent of the normal competitive price, even under assumptions most favorable to a tariff. Administrative costs for such a tariff could easily be so large that no tariff at all would be the best policy.

Since stockpiling is such a basic policy for a commodity with inflexible (inelastic) demand, we determined efficient tariffs on the assumption that optimal stockpiles were determined at the same time. The tariff generally reduced expected disruption costs by no more than $1 million in the various scenarios, although tariffs did allow efficient stockpiles to be decreased by between 3 and 6 percent. In general, stockpiles alone appear able to do almost as well as a combination of stockpiles and tariffs. In addition, stockpiles may cause fewer political problems than would tariffs, as the higher costs induced by a tariff would be very visible. It does not appear that tariffs on manganese are an effective method of guarding against supply disruptions.

The Benefits of Development of Technical Adaptations

Government policies could encourage technical changes to reduce manganese consumption, make demand more elastic, or speed the rate of adjustment to the long-run equilibrium under disruption conditions. Such policies can yield contingency benefits—reductions in expected disruption costs. These benefits will be in addition to any commercial benefits accruing during normal times. In all cases the contingency benefits must be weighed against the costs and other consequences of the policy. Because the range of possible projects is enormous, and because the costs are generally unknown, we have not attempted to evaluate specific projects. After presenting estimates of the contingency benefits associated with particular policies, we indicate in general terms the kinds of policy actions that might be taken.

Increasing the speed of investment during a disruption reduces other adjustment costs incurred by users who would otherwise have to use inappropriate equipment or production practices. Experiments with our optimal policy program suggest strongly that more rapid introduction of techniques such as external desulfurization and continuous casting during disruptions is not likely to be a cost-effective way of responding to supply crises.

In the case of the 15 percent disruption scenario, when efficient stocks are held, the contingency benefits of a 10 percent reduction in base-level consumption range from $48 million for the disruption probabilities of Scenario A to $129 million in Scenario C. For the 26 percent disruption, the benefits range from $75 million in Scenario A to $200 million in Scenario C. In a situation of a possible 26 percent disruption, with an inception probability of 0.2, the contingency benefits of a 10 percent consumption decrease are $137 million for a 0.5 continuation probability, but rise to $318 million for a 0.9 continuation probability. These calculations assume no deterrence effects of conservation in consumption, that is, reductions in U.S. consumption prior to a disruption do not affect the assumption that imports equal to 26 percent of normal U.S. consumption are lost during disruptions, or the assumption that disruption probabilities remain fixed at the values specified in Scenarios A, B, or C.

The likely duration of a crisis is important not only for the extent of damage inflicted, but also for the benefits which various countervailing policies can provide. If long disruptions are more likely, policies like technological adaptions, which yield sustained benefits and do not become depleted as do stockpiles, can yield benefits over a great period of time and be relatively more cost-effective. The contingency benefits calculated above are large because they are based on severely pessimistic scenarios.

Technological policies to promote the implementation of new technologies could increase the price elasticity of demand, that is, decrease the cost of efforts to reduce consumption. For a 26 percent disruption in Scenario B (a 0.2 inception probability and a 0.5 continuation probability), a 10 percent increase in the long-run elasticity gives very small contingency benefits of $2.1 million, or less than 0.2 percent of total expected costs. Increasing the short-run elasticity by 10 percent yields even smaller benefits.

Possible Demand-Side Policies to Reduce Crisis Impacts

What sorts of policies would be justified by the estimated contingency benefits? The benefits are probably unrealistically large, in part because the assumed scenarios are so pessimistic. Nevertheless, some technological projects appear justified.

A modest government program to gather and organize information on the control of iron and steel production practices, which could lead quickly to reduced U.S. manganese consumption, might well be justified. It would be desirable for such information to be made immediately available to all firms in the industry, especially if a manganese-supply disruption began to appear imminent. However, for policies which speed U.S. response to disruptions, it should always be asked whether the response time saved by a given government program could not be purchased more cheaply by simply having more inventories on hand at the beginning of a disruption. We have considered whether this would be best achieved by a government stockpile or by government encouragement of expanded private inventories.

Most of the currently known, important techniques for conserving manganese appear to have already been researched and are at least partially developed. Processes such as continuous casting and external desulfurization are likely to become standard practice in future years in any case. The most plausible type of government-sponsored technological preparation would simply involve documentation and improvement of existing technologies, such as accumulation of detailed engineering information on advanced ferroalloy-addition practices; however, these matters are being studied by private industry. As a general rule, the most obvious projects that would be candidates for government sponsorship would involve expensive basic research, the benefits of which would ideally accrue to all manganese consumers, not just to the firms engaged in the research.

As discussed above, timely adjustment in AISI steel specifications could significantly accelerate technological adaptations during an emergency. The government, working with the AISI and other technical experts, might be able to streamline and speed the process of specification change. Possible actions include convening panels to evaluate potential specification changes in light of available technical information. The private sector will normally undertake such activities, so the government program need not be massive.

Acceleration of Ocean Mining

By far the most interesting area in which U.S. policy can play a role toward developing technologies affecting the manganese market is in ocean-nodule mining. U.S. land-based manganese resources show little promise of becoming profitable, even during serious disruptions, unless they are very sustained. However, the contingency benefits of developing seabed mining could well be large enough to justify government subsidization. As more information about the economics of manganese recovery from ocean nodules becomes available, this question can be considered in detail.

The profitability of manganese recovery from ocean nodules is apparently uncertain at current prices. Information currently available is insufficient to assess, with any precision, what subsidy, if any, would be required to induce investment in manganese recovery. It may well be the case that simply guaranteeing access to the seabed for U.S. firms would induce commercial exploitation of the nodules. The contingency benefits of such an action could be substantial. There may also be ways to subsidize, or threaten to subsidize, manganese production from nodules without an unwarranted expenditure of public funds. The subsidies might be justified by the contingency benefits of reducing imports; such benefits would be similar to those of reducing consumption. Any subsidy program should concentrate on reducing risk, not simply on transferring funds. In addition, political problems of ownership should be resolved before any substantial expenditures are made.

As one possibility the government could sell price guarantees on deliveries of sea-nodule manganese to be made in specified years and in specified quantities. The guaranteed prices could be set moderately above estimated competitive market prices and could be specified in constant dollars. Private companies would bid on the guarantees, and payments from a winning company could be made in regular installments, with payments to cease whenever the purchaser wished to let the guarantee lapse. The purpose of having investors in ocean-nodule mining bid for the price guarantees would not primarily be to raise revenue. The resulting payments would only be large enough to demonstrate the producers' continued intentions to make the investments to produce the required manganese. The prices and quantities in the guarantee should be generous enough so that more than one company would be interested in bidding.

It would seem natural to give such subsidies only to U.S. companies, though much the same effect on the manganese market would be obtained if, say, a Japanese or Canadian producer were subsidized. Most current deep-sea mining ventures in fact involve international consortia. Perhaps governments of other manganese-consuming countries could be encouraged to cooperate in a program of the sort outlined here.

By giving companies producing manganese from undersea nodules the benefit of the chance that future prices will exceed the guarantee, the financial commitment of the federal government could be lessened. The main uncertainty that private manganese nodule miners would face with price guarantees would involve the future costs of production. Such costs are probably less volatile than market prices for a commodity like manganese, and generally more under the control of the individual company. Manganese could be delivered and stored in a relatively unrefined form if the necessary further processing could be performed on relatively short notice after a disruption has begun.

A price-guarantee proposal such as that outlined above would be quite flexible, in that it could first be used on a small scale as a threat to combat noncompetitive tendencies among manganese-producing countries, and it then could be implemented to encourage production of whatever quantities of manganese were deemed appropriate, perhaps on more than one occasion. Its main disadvantage is the rather long time horizon before the manganese market would be directly affected.

We do not believe that present risks in the manganese market justify large government technology programs. If risks do increase, some support could be justified. A price-guarantee program of the sort described above is more flexible than a straight subsidy, and more directly attacks the problem of reducing risk for a new technology. As with all subsidy programs, this one could evolve into a form of protection for inefficient and policy-dependent interests, and therefore it should be implemented only after the most careful evaluation of risks and benefits. If private consortia continue their efforts to recover or store manganese, and if risks are not made much more severe (as, for example, by the accession of highly radical anti-Western governments in South Africa and perhaps Gabon, combined with an Australian move to the left), such a program need not be implemented.

NOTE

1. In Charles River Associates, Policy Implications of Producer Supply Restrictions: The World Manganese Market (Cambridge, Mass.: CRA, December 1976), we consider recycling as a measure to reduce the demand for inputs of primary manganese.

PART IV

COBALT AND PLATINUM/PALLADIUM: CONCENTRATED MARKETS AND PROBLEMATICAL RISKS

Formal cartels are unlikely to form in the cobalt and the platinum-group metals markets, for they are already largely monopolized—cobalt, by Zaire; and platinum/palladium, by South Africa and the Soviet Union. A formal cobalt cartel would produce very small long-run gains for most producing countries, and overt Soviet-South African collusion in the platinum-group markets seems most unlikely.

Market concentration, however, poses substantial threats, particularly in the platinum group, where the two major suppliers may have political motivations to restrict supplies, or may rapidly shift policies. We found the situation in the two markets to differ substantially. The short-term disruption costs for platinum-group metals are lessened by the fact that the critical users generally employ the metals as capital goods, and carry large stocks in use, and hence could weather disruptions of over a year. Except for large stocks, non-Soviet and non-African supplies are quite poor alternatives, though Canada makes some contribution.

Fewer cobalt uses are as critical as the uses of the platinum-group metals, though most users do not have the cushion of very large stocks in use, as is the case in the platinum group. In a severe crisis, domestic cobalt production could eventually supply at least some of the market within a reasonable time, though it would be costly, and deep-ocean nodules are a potential alternative in the long run.

The future advent of ocean mining and the development of new supply sources indicate that in the very long run, the cobalt market may become less concentrated than it is now. As this happens, there will be less potential for severe disruptions engineered by Zaire or caused by events affecting supplies from that country. In the platinum group, substantial deconcentration appears less likely. Decreases in use and increases in substitution possibilities would have to be the main source of decreased risks. Because knowledge of substitution possibilities in catalytic and other uses is widely fragmented and often closely held, government efforts could help enhance long-run flexibility.

In both cobalt and the platinum group significant stockholding is justified. The probability of formal collusion among producers in both these markets is probably even less than in the chromium and manganese markets, but the concentration of supply in potentially unreliable areas—distant regions of Zaire, in the case of cobalt; South Africa and the Soviet Union, in the case of the platinum-group metals—suggests that other types of disruptions may be most important. Since the focus of this book is on deliberate producer supply restrictions, disruptions such as the 1977 invasion of Shaba Province in Zaire are, strictly speaking, outside its scope.

However, the consequences of such a disruption are so similar to one deliberately engineered by producers, and the appropriate preparatory economic policies are so closely related, that in fact, most of the following analysis remains relevant. Our calculations of efficient economic contingency stocks are based on disruption probabilities that could be illustrative of disruptions caused either by exogenous political events or deliberate producer decisions.

Do any other major materials markets pose similar threats? Canada is predominant in nickel production, but its interdependence with the United States is so great that a cutoff in supplies is virtually unthinkable. Cobalt and the platinum-group metals represent relatively unfavorable cases, not typical of the range of market situations for many minerals.

8
COBALT: IMPACT AND POLICY ANALYSIS

As indicated in the introduction to Part IV, the potential for severe and damaging restrictions in the cobalt market due to deliberate actions by producers is problematical. While U.S. import dependence is virtually complete, except for releases from the government stockpile, the market value of 1974 consumption was only $65 million. As discussed in Chapter 3, the market is substantially monopolized by Zaire; while a cartel of African producers is feasible, the incentives for formation are very small at best. Our detailed analysis of the market showed that, unlike the pre-1973 situation in the oil market, even a totally effective cartel would have only moderate effects on cobalt prices in the long run; such a cartel action would not, in any sense, precipitate a long-run crisis for the United States or other consuming nations. On the other hand, Zaire is a dominant producer, with close to a 70 percent share of non-Communist production. If it chose to, Zaire could, for a time, impose a substantial embargo on cobalt supplies that would be very costly to the non-Communist world. Events of 1977 and 1978 indicate that the most relevant type of disruption to consider in the case of cobalt may be in fact, one due to exogenous political events. During earlier disorders in Angola, cobalt shipments had been substantially reduced and delayed, although the production flow had continued. However, growth of production had been relatively uninterrupted during the disorders in the 1960s, in what was then the Belgian Congo.

A cobalt-supply disruption would be costly for the United States, but at least some response is possible. In the short run, industry and government stocks could offset moderate supply interruptions. Over somewhat longer periods, some domestic land-based production could be resumed, supplying a portion of normal domestic demand, albeit at elevated prices, and shipments from alternative foreign sources could increase. Over longer periods and in the future, cobalt from deep-sea nodules could dramatically lower import dependence.

The demand side also constrains possible impacts. Unlike energy, cobalt is neither used ubiquitously nor is it critical to consumption by most individuals or to the processes of most firms. Unlike manganese, which is a vital input into steel and so into much industrial activity, cobalt is crucial to relatively few sectors of the economy. Chromium is virtually without substitutes as an ingredient of stainless steel, which itself is highly important to a number of processes. Over the intermediate to long run, moderately good substitutes are available for at least some cobalt end uses. Sudden and severe supply disruptions would cause short-run dislocation and hardship, and shifts to alternatives would take time and entail costs; in some end uses, sudden and severe supply restrictions could force costly curtailments and lead to manyfold price increases.

OVERALL POLICY CONCLUSIONS FOR COBALT

The potential likelihood and possible impact of a profit-seeking cobalt cartel appear too small to justify significant investments in contingency policies directed at blunting a cartel movement. On the other hand, Zaire's dominant market position makes embargoes possible. For example, in periods following prosperity in the copper market, Zaire could tolerate some loss in current foreign exchange earnings. Therefore, policy might be designed to assure a reasonable proportion of consumption during periods of supply disruption, though Zaire's motivation for such disruption (under present conditions) is unclear. Subsidies for domestic production prior to disruption do not appear desirable because of the high domestic costs of output. Similar factors militate against the use of tariffs.

On the technological side, relatively large efforts appear unwarranted, as substitution possibilities are reasonably well known in many areas. Modest research efforts to develop substitutes for cobalt superalloys (such as high-temperature, high-strength ceramics), and information centralization and evaluation would be helpful.

On the supply side, limited technological efforts appear desirable. Efforts to ensure the viability and accessibility of technologies to utilize domestic ores could yield some benefits. Potentially more significant, the government might encourage efforts to perfect ocean-mining technology. Such materials policies affect diplomatic considerations, as viable ocean mining requires prior resolution of questions concerning both property rights and international environmental policy. The nodules also contain nickel, copper, and manganese, as well as a number of other minerals in small concentrations. The prospective contingency benefits of nodule development relative to copper appear small. But, as discussed in Chapter 7, the manganese-market risks are substantial, and manganese-market prospects might justify price supports or other policy actions, thereby making nodule development a more promising and less costly policy for guarding against contingencies in the cobalt market.

In the remainder of this chapter, we briefly survey demand and supply options, present estimates of the possible impacts of cobalt cartels and embargoes, and evaluate the policy options discussed above.

DEMAND-SIDE OPTIONS: SUBSTITUTES AND ALTERNATIVES

Because the structure of the cobalt market would permit Zaire to disrupt supplies severely, it is important to look at substitution and replacement possibilities. The examination discloses potentially critical areas where research might be beneficial.

Most of the possible flexibility (elasticity) of cobalt demand comes from substitution, not from shifts in final product choice. To determine substitution possibilities, we performed both quantitative and qualitative analyses. The results of these analyses were consistent: substitutes are available for a number of uses, and superalloys represent one of the less flexible of the major end uses. Most substitutions for cobalt-bearing materials are either relatively costly or involve a degradation in performance.

To analyze the demand for cobalt, it is useful to divide its consumption into the following four major categories: superalloys; magnetic alloys; other alloys; and oxides, salts, and dryers. Though there are year-to-year variations, consumption is about equally divided among these categories.

Superalloys

Superalloys are used principally in the construction of gas turbines. Aircraft engines constitute by far the greatest use of gas turbines; other uses include electricity generation, natural gas pumping, and marine propulsion. There are four types of superalloys: iron-based alloys with chromium and nickel; complex chromium-nickel-cobalt-iron alloys; nickel-based alloys; and cobalt-based alloys. Each group of alloys has particular attributes that make it most appropriate in any given situation. However, cobalt- and nickel-based alloys generally have similar performance characteristics. Cobalt is typically used in all of the superalloy types, but other metals can be substituted for cobalt to a limited extent. In particular, nickel and tungsten are somewhat substitutable for cobalt in jet engine production. Because the value of the cobalt content represents about 0.3 percent, or less, of the value of a jet engine, the price effect of cobalt on the manufacture of jet engines is negligible (unless cobalt prices were to increase by several orders of magnitude). Because cobalt is much more expensive than nickel, the price of cobalt has a significant effect on the choice of superalloy. However, cobalt-based superalloys do have some production and performance advantages over nickel-based superalloys. They can be melted in air, whereas nickel-based superalloys generally must be melted in vacuum, and they are more easily welded and hence repaired.

The price elasticity of cobalt demand in superalloys is less than for other major end uses: about -0.03 in the short run and about -0.4 over longer periods. Though substitution possibilities exist with current technology, substitution can take as long as five years, and it is likely that very sharp declines in the availability of cobalt to superalloy manufacturers would cause significant short-run disruptions. For this reason, priority might be attached to ensuring stable supplies of cobalt for consumption in superalloys. On the other hand, superalloy manufacturers might be willing to pay a much higher price than would other consumers, at least in the short run. Over the very long run, probably beyond 1990, the use of gas turbines in automobiles and trucks could increase demand in this end use and reduce the overall price elasticity of cobalt demand.

Magnetic Alloys

Cobalt is used in combination with nickel and aluminum to form alnico magnets. New developments in magnetic alloys may make cobalt a more important metal in this area. There are many types of alnico alloys in which cobalt content typically varies between 8 and 24 percent. Nickel is somewhat substitutable for cobalt in such magnets. The major use of magnets is in speakers for radio, television, and sound-reproduction equipment, and there is a substantial substitution potential between alnico magnets and ceramics in this use. Often, as in the case of television, for example, certain qualities of alnicos make them preferable to ceramic magnets; alternatively, ceramic magnets, by virtue of their lower price, are more common in other uses, such as phonographic equipment. Demand is relatively price elastic (about -0.7 in both the short and long run) for this end use, owing to the substantial opportunities for substitution with known technology.

Other Alloys

Cobalt is used in a number of metallic alloys typically used for their heat resistance and hardness. Some amounts of cobalt enhance stainless, heat-resisting, and wear-resisting steels. Nonferrous alloys with cobalt are used in welding and hard-facing rod materials and in cutting and wear-resistant alloys, primarily cemented carbides. The proportions of cobalt within an alloy can vary, and the types of alloys themselves often can be interchanged for the same end use. Substitute materials include steel, molybdenum, tungsten, and nickel. Because of the substitution possibilities, the demand for cobalt for these end uses is moderately price elastic.

Oxides, Salts, and Dryers

Cobalt consumption in oxides, salts, and dryers includes that cobalt which is used in pigments, dryers, catalysts, electroplating, and pharmacological and diet supplements. Substitution possibilities in dryers exist. Little is known about the substitution potential between cobalt and other materials in catalysts, although it is expected that there will be substantially increased demand for cobalt catalysts in the future, owing to air-quality regulations. In electroplating, cobalt is substitutable with nickel, but other substitution possibilities are, at this time, unknown. Cobalt use in pharmaceuticals and diet supplements appears to be quite inflexible, but this end use is relatively minor. All things considered, the demand for cobalt in this group of end uses is probably at best moderately price elastic.

General Consideration of Substitutes for Cobalt

Table 8.1 presents a list of substitutes for cobalt in decreasing order of the potential for substitution. The table should be interpreted with caution insofar as this ordering represents qualitative and somewhat speculative judgments. Nonetheless, this list can be used as a guide in examining the availability of substitute materials and the speed with which demand adjustments can be made.

The availability of alternatives to cobalt is an important consideration in determining the severity of potential cobalt-supply disruptions. Nickel and iron are the two most important substitutes for cobalt. Both are exceedingly abundant as compared to U.S. cobalt consumption, and supplies of both are relatively secure. The next two potential substitutes, tungsten and chromium, are also abundant relative to U.S. cobalt consumption. As discussed in Chapters 3 and 6, chromium supply is not necessarily secure; tungsten supply is a matter of somewhat less concern, though imports are substantial.[1] The rest of the substitute materials are abundant domestically.

TABLE 8.1

Materials Ordered by Degree of Substitutability for Cobalt

Substitute Material	End Use for Cobalt
Nickel	Superalloys, magnets, other alloys, nonmetallic
Iron	Superalloys, magnets, other alloys
Tungsten	Superalloys, other alloys, nonmetallic
Chromium	Superalloys
Molybdenum	Superalloys, other alloys, nonmetallic
Aluminum	Magnets, nonmetallic
Ceramics	Magnets
Copper	Magnets

Source: Charles River Associates, Policy Implications of Producer Country Supply Restrictions: Overview and Summary (Cambridge, Mass.: CRA, 1976), p. 194.

The more rapidly substitutes can be implemented, the less severe will be the impacts of a supply disruption. Our research indicates that in a number of end uses, the demand for cobalt is much less elastic over short than over long periods. This means that the costs of shifting quickly are substantially higher than the costs of allowing more time for adjustment. Stocks can provide a valuable cushion against the sudden need to rely on costly short-run substitution possibilities. Little concrete information is available on the speed with which substitutes can be implemented, as there had been no recent crises at the time of our research.

A 1969 crisis in the nickel market provides clues that rapid adjustment is feasible in some areas. In 1969 a protracted strike by Canadian mineworkers caused a severe shortage of nickel in the United States, and prices of nickel rose by factors of up to six. Within weeks, it was found that cobalt could be substituted for nickel in electroplating to a much greater extent than had previously been realized. The new applications were developed over a short time period, and were not the results of applying off-the-shelf technology. After the strike ended, users shifted back to nickel. These short-term possibilities are still available and may be indicative of possibilities in other areas.

Cobalt supplies generally were quite reliable until the mid-1970s, so users have had relatively little incentive to investigate possible substitutes. Our quantitative and qualitative work indicates that in many uses, while short-run substitution is difficult, substantial long-run substitution is possible.

Effects of Supply Disruptions on Cobalt Users

If no cobalt were available, short-term economic dislocations would be severe. Thus, it is important to ensure the availability of supplies, equaling at least some fraction of current consumption, for a prolonged disruption.

Priority end uses for cobalt include superalloys for jet turbines and other alloys used in industrial production, such as cemented carbides. Magnetic alloys and nonmetallic uses appear to have more substitution possibilities, though cobalt for medical purposes (while a negligible end use in terms of quantity of cobalt) would also have a high priority. Cobalt for the priority end uses includes about half of current U.S. consumption. Substitution among alloys in these end uses in a severe disruption would allow cobalt consumption in them to be cut by on the order of one-third. If cobalt consumption in the other uses could be cut by two-thirds, then during a severe disruption, cobalt consumption could be reduced by 50 percent.

As a result, very severe short-term economic dislocations need not occur until cobalt supply declined to one-half of normal levels. In the long run, technological development of new alloys that serve the purposes of current cobalt-containing alloys would allow the United States to be even less dependent on cobalt. Relatively high costs would be incurred, but trauma need not occur.

SUPPLY-SIDE OPTIONS: SHORT-, MEDIUM-, AND LONG-TERM ALTERNATIVES

Present government stocks of cobalt are large, and domestic land-based resources are costly but plentiful. During World War II U.S. production expanded significantly. Undersea nodules could contribute large secure supplies. In addition to the domestically secure supplies, other industrial countries, such as Australia, Canada, and Finland, are unlikely to participate in cartels or supply cutoffs; these countries have substantial production and resources, although cobalt usually is produced as a byproduct of other mining operations. Finally, New Caledonia, the Philippines, and South America contain huge, largely unexploited resources. All of these sources could expand in the event of a cartel action or supply cutoff, if it were long enough.

Domestically Secure Supplies: Actual and Potential

We now consider domestically secure supplies in order of the speed of response they would offer in the event of a cartel action: stocks, secondary recovery, land-based deposits, and ocean nodules.

Stocks of Cobalt

In the short run, existing stocks of cobalt could be used in the event of a supply cutoff. The exact amount of domestically, privately held cobalt is not known, but was probably in the 2-to-3-million-pound range in 1976 or several months' supply. The amount of cobalt in the General Services Administration strategic stockpile stood at about 40.8 million pounds as of July 31, 1976, or more than twice current annual consumption. However, the stockpile objective announced in October 1976 was about double the existing stockpile. As a result, government cobalt stocks may well not be readily made available for civilian purposes in a nonmilitary supply crisis.

The government stockpile could be a substantial deterrent to any attempt by foreign suppliers to embargo cobalt supplies, if available for such purposes. By the time such a stockpile would be exhausted, consumption would have been reduced, and, as we will see below, other supply adjustments would probably be well advanced. The existence of the stockpile may also moderate producer attempts to raise cobalt prices. Calculations based on computer simulations of our econometric model of the world cobalt market show that stockpile releases have significantly depressing effects on price. Following evaluation of the other supply alternatives, we explore the question of the most appropriate size for a cobalt stockpile.

Secondary Recovery

The major sources of recycled cobalt currently are the grindings and turnings generated in cobalt fabrication. The potential for cobalt recovery from used products has not been realized, apparently because cobalt content is typically only a small proportion of the fabricated goods. Data on the current rate of secondary recovery are unreliable, but indicate annual supplies of as much as 1 million pounds.

The capability for expanding secondary recovery probably exists. During the Korean War, cobalt was under allocation, and domestic recovery for domestic uses increased from around 20,000 pounds per year to 1.5 million pounds per year. (This increase may reflect anomalies in the data to some extent.) The sources of such recovery were jet turbines, the blades of which have a relatively short life (around three years), and permanent magnets, which have a much longer life span.

Though secondary-recovery efforts for cobalt can be expanded rapidly in the event of a crisis, the potential annual supply from such activity is limited. It is unlikely that more than 20 percent of annual consumption would be accounted for by scrap over the long run. This would, however, make a substantial contribution toward meeting priority demands.

Domestic Land-Based Production

Domestic deposits are not being mined currently. Prices that are expected to remain substantially above historic levels would be required to induce significant production. The total scope of domestic cobalt resources is not known with much certainty, as neither prices nor the size of the market has provided much incentive for exploration. As of 1962, the following resources were known as available:

Property Owner	Amount of Resources
Bethlehem Cornwall Corp., Pennsylvania	56 million pounds of contained cobalt
Calera Mining Co. Idaho	200 million pounds of contained cobalt
National Lead Corp., Missouri	16 million pounds of contained cobalt

Total resources are much greater; a recent Bureau of Mines estimate is 1.68 billion pounds, which is also an understatement, as the massive copper-nickel deposits near Ely, Minnesota, have yet to be explored fully. These deposits contain at least 1 billion pounds of cobalt, and may contain much more.

Moderate price increases of up to about 20 percent probably would not induce mining of most U.S. deposits. Long-term cartel price increases toward the top of the plausible range—about 50 percent—might induce production from the Calera property, but not from the others.

Research into the cobalt market during World War II shows that domestic mining and refining can expand rapidly. Between 1939 and 1941, domestic mining increased from zero to 523,000 pounds per year (6 percent of world production); by 1945, production was 1.28 million pounds per year (12 percent of world production). During this period, refining capacity increased substantially.

From the World War II experience and other factors, it appears that in the event of a supply cutoff or other severe crisis, the United States could conceivably generate relatively quickly about 3.5 million pounds per year in capacity from domestic deposits, which is about 20 percent of consumption rates in the mid-1970s. Such an increase in production capacity would require private firms to form an expectation that the crisis is likely to persist for a number of years. Refining capacity would be able to adjust to even greater output.

Deep-sea Manganese Nodules

In the past several years, commercial interests have made significant investments to evaluate the feasibility of exploiting deep ocean deposits of manganese nodules. The metals of interest to the ventures include nickel, copper, and cobalt.[2] Details of the mining technology are closely guarded, but it appears that most major technical problems have been solved. There is, however, substantial uncertainty about the cost of operation. Preliminary estimates indicate potential profitability under normal economic conditions.

Political factors appear to be the main deterrent to ocean mining, since companies are not assured property rights in nodule-rich portions of the sea bed. Questions of national versus international ownership and control have yet to be resolved. The severe conflicts among resource-rich and resource-poor countries, among the land-based and the maritime countries, and among the industrial and the less developed nations indicate that these problems will not be resolved easily. It is therefore likely that ocean mining will be delayed well beyond the time at which it becomes technologically and economically feasible.

The potential of these nodules is vast. Initial estimates of cobalt content indicate an amount of about 250,000 times 1973 annual U.S. consumption. A small mining operation might yield 4.8 million pounds per year of cobalt; four such operations would have satisfied U.S. demand at 1973 levels and would have been one-third of total non-Communist-world production. Such output would undoubtedly depress cobalt prices; our estimates indicate that moderate-scale mining could depress prices by between 15 and 39 percent, depending on the elasticity of cobalt demand. At present cost estimates, this need not make ocean mining uneconomic. The extent of this mining could effectively prevent any potential cartel from elevating prices, and it could also destroy the possible usefulness of the embargo weapon.

Potential Foreign Supplies

To complete the inventory of supply alternatives, we note that large low-grade deposits in New Caledonia now are being investigated by French and other commercial interests. Plans in 1975 were for 1980 output of over 22 million pounds per year, but it is clear that these plans will not be realized, at least by such an early date. It is not likely that New Caledonia would join a cartel of African producers. New Caledonian output would not only challenge any cartel attempt to raise long-run prices, but could challenge Zaire's dominance of the market and reduce prices.

IMPACTS AND POLICY IMPLICATIONS OF POTENTIAL CARTEL ACTIONS

As discussed in Chapter 3, even a complete cartelization of foreign cobalt supplies would yield only small revenue and profit gains to producers in the long run, although short-run gains could be substantial. A cartel including Zaire, Zambia, and Morocco might be feasible, and would control over 80 percent of recent non-Communist output. Complete cartelization is most unlikely, as the remainder of output comes from developed countries, which would be unlikely to restrict supplies. We established an upper bound on cartel effects by using our computerized model of the market to simulate the effects of a complete cartel. We found that such a restructuring of the market would raise long-run prices by between 17 percent and 50 percent.[3] In addition, a total cartel is most unlikely, so our calculations may overstate the likely degree of market power.

Based on these price effects, our estimates of the annual cost to the United States of a cartel price increase, using 1973 prices and consumption, were as follows:

Price Increase	Annual Cost
16 percent	$ 9.4 million
50 percent	25.3 million

These costs include increased transfer payments to foreign producers for remaining imports, and losses to users because of shifts to more expensive or less effective alternatives.

It thus appears that a cobalt cartel oriented toward its long-run economic interest would neither raise prices precipitously nor impose large costs on the United States. Taking the high cost estimate, if consumption were to grow at a high rate of 5 percent per year, and if the cartel were to endure and charge a high monopoly price, at a 10 percent discount rate, the total present-value cost to the United States would be on the order of $125 million.

General Policy Implications

Because the likelihood and potential impact of a cobalt cartel are not disastrously high, major cartel-oriented policy initiatives do not appear warranted. This is unlike the case of bauxite, in which the cost of an IBA high-price policy could justify substantial efforts. Cobalt prices would be likely to rise by, at most, 50 percent in the long run, at which price most of the domestic land-based resources would still be of doubtful economic viability. Hence, subsidies for production could not yield contingency benefits in the event of a plausible cartel action. Subsidies also might have to be substantial, with the usual redistributive effects on income. Tariffs would be similarly inefficient on the supply side, though they could induce significant economizing by users. We cannot recommend their use, as it appears that short-run problems can be best handled by stocks. Adjustments could be achieved quickly, and long-run cartel success is sufficiently improbable that substantial tariff-induced distortions of production and consumption do not appear justified.

Cobalt-Embargo Possibilities: Potential Impact and Policy Implications

We now consider embargo possibilities and impacts, and then evaluate policy options, particularly stockpiling. Much of this analysis is also relevant to a disruption scenario in which, for example, Zaire's copper- and cobalt-producing regions are threatened by invasion by rebels from neighboring Angola, as in fact happened in 1977 and 1978. For convenience, and because deliberate restrictions were our main focus, the following discussion of contingencies is predominantly phrased in terms of deliberate supply restrictions imposed by Zaire. Such disruptions appear less likely than those due to exogenous political forces.

The dominance of Zaire suggests that the possibility of an embargo imposed for political reasons may be relatively high. Zaire depends heavily on copper for its foreign exchange, while Zambia, the second largest producer, depends almost exclusively on it. In addition, if copper prices were low, Zaire might be tempted to increase the cobalt price in order to reap short-run monopoly profits and trade lower foreign exchange earnings in the future for higher ones in the present.

Consuming countries might counter an embargo or price escalation by restricting copper purchases. The United States at least in principle could survive on stocks (although GSA stocks might well not be available during an economic emergency), and alternative supplies could expand rapidly. An embargo would induce consumers to search for and implement substitutes. Political upheaval in Zaire could result in longer-term embargoes or supply interruptions.

Our examination of substitution possibilities showed that the total unavailability of cobalt could cause severe dislocations, particularly in sectors using superalloys that contain cobalt. But these priority needs could be met with half, or less, of normal supplies.

In the event of a cobalt-supply cutoff, the United States has nearly three years' supply in current stocks, including strategic stocks. About 25 percent of U.S. cobalt consumption could in principle be supplied from land-based reserves, and about 15 percent can be accounted for by secondary recovery. These activities would extend immediate domestic supplies for about five years, by which time U.S. demand could be supplied by crash programs in undersea mining and Minnesota sulfide extraction. If strategic stocks were not available during peacetime emergencies, the implication is that a contingency stockpile of comparable size might be warranted.

Analysis of Embargo Scenarios

Since the current market structure could permit severe supply disruptions, we used our optimal policy models to evaluate impacts and policies. The scenarios analyzed are all quite pessimistic. We assumed a total cutoff of supplies from Zaire, and imports from other countries were assumed constant at their 1974 rates. Domestic production was assumed to begin at a price about 50 percent above the 1974 level, and potentially to attain a peak capacity of about 3.3 million pounds. In addition, this production was assumed not to begin unless the cutoff was expected to be lengthy. Though severe embargoes are unlikely, we analyzed three pessimistic scenarios that set bounds on the possible impacts and, hence, on the desirable extent of policy actions. The cobalt-embargo scenarios are as follows:

Scenario	Probability of Occurrence in Any Given Year	Probability of a Continuation for Another Year
A—short-term disruption	.10	.10
B—long-term disruption	.10	.90
C—very likely moderate-term disruption	.30	.40

In all scenarios, supplies from Zaire are assumed to be totally cut off.

For case A, the scenario assumptions mean that the probability that a disruption will occur in any given year is 0.1, and that, if the disruption occurs, the probability that it will continue for another year is also 0.1; the probability that it will continue for a third year is 0.01; for a fourth year, 0.001, and so on. The inception probabilities imply that the probability that a disruption will occur within five years is 0.41, and within ten years, 0.65.

Scenarios B and C provide alternative upper bounds to disruption likelihoods. Case B assumes an initial probability of 0.1 and a continuation probability of 0.9. This corresponds to a potential long-run disruption with a high initial probability; it implies that the probability that a potential long-run disruption will occur within five years is 0.41, and within ten years, 0.65. The probability that the disruption will last longer than five years is 0.59. Case C assumes an initial probability of 0.3 and a continuation probability of 0.4. This corresponds to an expected disruption with a very high initial probability; the implied probabilities that a disruption will occur within five and ten years are 0.83 and 0.97, respectively. The probability that the disruption will last longer than five years is only 0.01.

Using our forecasting model of the world cobalt market, we studied the effects on the cobalt price of short-run cutoffs in supply from Zaire. Our analysis showed that a cutoff of only six to nine months would lead to a tenfold, or greater, increase in price if stocks were not available.

Policy Options for Guarding against Possible Disruption of Supplies from Zaire

The estimates of economically efficient contingency stocks for the three pessimistic scenarios (assuming no deterrence effects) discussed above are as follows:

Scenario	Contingency Stocks in Million Pounds	Equivalent in Months of Consumption
A—short-term disruption	13	8
B—long-term disruption	32	20
C—very likely medium-term disruption	27	17

Scenarios B and C are quite pessimistic and probably can be regarded as upper bounds to the efficient levels of stocks without deterrence effects. Of course, if advance knowledge of a disruption (such as that caused by the 1978 invasion of Shaba Province in Zaire) were available, much larger stockpiles could be justified; however, increasing a stockpile tends to be difficult after a disruption becomes more likely.

As discussed further in the Appendix to this volume, including deterrence effects in an analysis can lead to significantly higher estimates of efficient stock levels. However, in the cobalt market, deterrence effects for disruption probabilities are unlikely to be important, and price-deterrence effects for prices alone only increased recommended stockpiles by 30 or 40 percent in experimental runs with our generalized policy model.

Other policy options for ameliorating the cost to the United States of disruption in the cobalt market are much less promising than basic stockpiling. As noted above, some research may be useful to improve substitution possibilities in the superalloy area. If political changes or turmoil indicated that Zaire is likely to become an unreliable supplier, the government could consider research and policy action, particularly in the diplomatic area, to accelerate the development of ocean mining. Limited research on the technology required to mine domestic land-based deposits might be beneficial, as similar deposits are being exploited elsewhere.

NOTES

1. See Charles River Associates, Economic Issues Underlying Supply Access Agreements: A General Analysis and Prospects in Ten Mineral Markets (Cambridge, Mass.: CRA, July 1975), app. X.

2. It appears that ocean mining will have relatively minor impacts on the copper market. See Charles River Associates, Policy Implications of Producer Country Supply Restrictions: The World Copper Market (Cambridge, Mass.: CRA, December 1976). The manganese-market implications are potentially more significant, and are addressed in Chapter 7 of the present volume. See also Charles River Associates, Policy Implications of Producer Country Supply Restrictions: The World Manganese Market (Cambridge, Mass.: CRA, December 1976).

3. The range is determined by a range of alternative estimates of demand elasticity. See Charles River Associates, Policy Implications of Producer Country Supply Restrictions: The World Cobalt Market (Cambridge, Mass.: CRA, December 1976).

9
PLATINUM/PALLADIUM: IMPACT AND POLICY ANALYSIS

The major platinum-group metals—platinum and palladium—are among the most important raw materials for which the United States is and may remain virtually totally import dependent. In 1974 the value of sales of platinum and palladium to primary producers was over $280 million, larger than most of the markets included in this study. Petroleum refining and chemical production are particularly dependent on platinum metal catalysts. Catalytic converters, a major tool for control of automotive pollution, depend on platinum and palladium catalysts, as do a number of other pollution-control devices. Thus, platinum-group metals are important not only for continued industrial growth, but also for the achievement of environmental-quality goals. In this study we examined the implications of U.S. import dependence on platinum and palladium. While the remaining platinum-group metal markets are important but substantially smaller in size, an interruption in platinum-group supplies from South Africa could have severe effects in these markets as well.

In 1975, South Africa accounted for 64 percent of world platinum supply and 23 percent of world palladium supply; the Soviet Union accounted for 28 percent of world platinum supply and 69 percent of world palladium supply. Most of the world's remaining supply came from Canada. There is no indication that this near total dependency on South Africa and the Soviet Union for their supplies of platinum-group metals will decline in the immediate future, although it does appear that South Africa and the Soviet Union possess sufficient resources to sustain current rates of supply indefinitely.

Supplies of platinum and palladium to the United States and the rest of the non-Communist world are obviously extremely insecure.

Platinum and palladium are already effectively cartelized; it does not appear that substantial restrictions in sales would be in the economic self-interest of South Africa and the Soviet Union. However, political events in the Soviet Union and South Africa could well lead to the total cessation of exports from either or both countries. While prediction of such noneconomic events was beyond the scope of our research and expertise, the potential consequences of an interruption in platinum and palladium supplies (as well as supplies of other platinum-group metals) cannot be ignored by policy makers.

The consequences of an interruption in platinum and palladium supplies would be severe, but need not be traumatic, as users in most critical end-use categories employ platinum-group metals as capital goods and hold large stocks in use; U.S. users could generally withstand disruptions of at least one to two years, with only moderate cost escalations. In addition, users and dealers hold large metal stocks, roughly equal to a year's consumption. GSA stocks equal roughly a half year's consumption, but strategic goals were raised sharply in late 1976 to considerably more than a year's consumption. Short-run substitutes exist for many end uses, and over 30 percent of world platinum consumption is accounted for by jewelry purchases in Japan. In the event of a supply disruption, these uses would probably decline, as supplies would be bid away to more critical end uses. Increases in secondary recovery would also moderate disruption impacts.

GENERAL POLICY RECOMMENDATIONS

We evaluated tariffs, subsidies for the development of domestic supplies, stockpiles, and technological incentives as means of protecting the United States against the possibility of a supply disruption. Tariffs of moderate size are unlikely to increase domestic supplies, though they would reduce consumption somewhat.

Subsidies for domestic development are not likely to be effective in addressing the overall problem of U.S. production. For many years the United States maintained a subsidy program, but it met with little success. The major problem is that domestic deposits of platinum metals, except as minor constituents of some copper ores, are rare. Recently, deposits have been discovered in Montana, but at present, these are regarded as unlikely to meet a large percentage of U.S. needs, even in the long run and with high subsidies.

The existence of stocks in use and the potential for increased recycling could enable such critical users as the petroleum industry to continue operating in the presence of a short-run reduction in new metal supplies. If a disruption were protracted, new facilities would tend to use processes less dependent on platinum-group catalysts. The principal destabilizing influence during an emergency would probably be speculative stockholding, which would be encouraged by the role of platinum as a store of wealth, as an alternative to gold. Speculative hoarding in an emergency could lead to enormous increases in platinum-group prices, resulting in the diversion of supplies from industrial users. The role of government stocks is therefore twofold: to moderate the effects of speculatory hoarding (assuming more direct measures are not taken); and to ease the transition to nonplatinum-based technologies, should the disruption last indefinitely.

Because the effects of speculative hoarding are impossible to predict with any precision, it is difficult to reach conclusions about the appropriate magnitude of contingency stocks. Clearly, substantial stocks may be warranted. If speculative hoarding were not a factor, the capital-goods nature of platinum and palladium would lead to a conclusion that significantly smaller stocks are more appropriate than, say, for chromium and manganese. However, speculative hoarding would almost certainly be an important factor during an emergency, and a stockpile equaling as much as several years' consumption might well be warranted.

The number of platinum-using catalytic processes and the number of electrical applications are so large, and information is so closely held, that it is difficult to evaluate the possibilities for government technological policies to guard against the costs of an embargo. The most critical and vulnerable use of platinum metals is in catalytic processes where alternative technologies do not seem readily available; there is clearly a need for a greater flow of information about existing and alternative technologies. The government might encourage the gathering and evaluation of information, although such efforts will be difficult because so many of the most promising substitution possibilities are closely held trade secrets, and industry cooperation may be difficult to secure. Alternatively, the government might conduct limited research into nonplatinum catalysts or engage in joint projects with industry.

Substitute technologies—albeit higher-cost technologies—exist for many of the critical platinum and palladium uses. The major problem during a supply emergency would be sustaining production at plants designed to use platinum- and palladium-based processes. The capital-goods nature of these metals and the potential for recycling, drawing down of stocks, and diversion from less critical uses suggest that production would not be severely affected. New plants constructed during an emergency that was expected to continue indefinitely could, in most cases, be designed to use nonplatinum- and nonpalladium-based processes. As a result, while government sponsorship of some basic research and development may be justified because of the public-good nature of such research, a crash program using substantial government expenditures to encourage substitution away from platinum/palladium does not seem justified.

In the remainder of this chapter, we summarize the basis for these conclusions, first discussing alternative sources of platinum and palladium, and then examining the various end uses and possible U.S. responses to disruptions. This examination indicates the most promising areas for research and technology incentives. Using this information, we then examine a range of policy options.

SECURE SOURCES OF PLATINUM AND PALLADIUM: ACTUAL AND POTENTIAL

We now examine secure sources of platinum and palladium in decreasing order of their reliability as sources to the United States and of the sources' responsiveness to disruptions or price escalation: stocks, secondary recovery, primary domestic production, and increased foreign production.

U.S. Stocks of Platinum and Palladium

The most secure supplies of new platinum are stocks held by U.S. refiners, importers, dealers, and users, as well as U.S. General Services Administration stockpiles. The limited data indicate that private-sector stocks in the mid-1970s were equal to over one year's consumption, not including stocks in use as capital goods in the more critical end-use categories. In addition, in mid-1978 GSA stocks equaled roughly a half year of U.S. platinum consumption and a year of palladium consumption; however, upward revisions in the strategic goals for both of these materials in late 1976 left deficits in their respective stockpiles approximately equal to a year's consumption of each material. The United States, as a whole, thus possesses stocks which would be an excellent buffer against relatively long-run disruptions if they were available for economic emergencies (which they are not).

Secondary Recovery and Recycling

Secondary recovery and recycling contribute over 15 percent of total U.S. consumption. Platinum metals are often recoverable because they resist corrosion, and many of their uses are nondissipative; industry sources estimate that 75 to 80 percent of scrapped platinum is salvaged. The importance of recycling could increase during a disruption as materials were turned over more rapidly. Our research into individual end uses indicates that recovery could be accelerated, adding to domestic supplies.

Domestic Primary Production

The United States has no prospects for becoming self-sufficient in the primary production of platinum-group metals. Maximum primary production was achieved in 1966 with an output of 52,661 troy ounces, approximately equal to 3 percent of U.S. consumption that year. U.S. production may increase in the near future if tests at the Stillwater Complex in Montana turn out well, although even then, it appears doubtful that domestic production will add significantly or rapidly to U.S. supply. A recent Bureau of Mines report estimated that at prices double their current levels, only 35,000 troy ounces would be produced by 1985, and only 30,000 more ounces would be produced at prices five times current levels. The U.S. Department of the Interior reported: "U.S. primary production can only be regarded as an important supply alternative 10 years from now and at prices much higher than current prices."[1]

The sparseness of the resource base means that subsidies for domestic production would be relatively ineffective. From 1950 to the early 1960s the United States offered a 70 percent subsidy for the development of new domestic sources; only one contract was ever issued and there were no apparent results.

Production from Foreign Sources

The major sources for platinum and palladium, South Africa and the Soviet Union, cannot be considered secure. The Soviet Union is dominant in palladium, while South Africa is dominant in platinum. If one country's supplies were disrupted, increased exports by the other could supply some of the deficit. For example, South Africa's Rustenburg Mines generally holds very large inventories of both platinum and palladium. However, such supply responses should not be considered likely; if South Africa's supplies were disrupted, it is reasonable to assume that the Soviet Union would try to take advantage of the situation and exercise pressure by restricting its sales. However, the 1974-75 experience in the palladium market suggests that such efforts may fail: users might simply sell off stocks, moderating prices on the world market.

Canada is the only other significant supplier, and its production is relatively insensitive to price, being a byproduct of nickel. While Canadian supplies, which equaled 6 percent of recent U.S. consumption, may be considered relatively secure for the United

States, they are unlikely to increase dramatically. During a severe crisis the United States might be able to increase its access to Canadian supplies. There are no other known actual or potential sources of significant supplies of platinum-group metals.

THE USES OF PLATINUM-GROUP METALS AND MARKET RESPONSES TO SUPPLY DISRUPTIONS OR PRICE ESCALATIONS

Except for stocks and secondary recovery, supply alternatives for platinum-group metals appear extremely limited. Because severe disruptions are possible, the alternatives for end-use sectors become critical. In this section we examine the positions of the major end-use categories: petroleum, chemicals, and automobiles, which are mostly catalytic end uses; electrical, glass, medical, dental, and jewelry end uses, which are noncatalytic. We first consider a fundamental distinction—between the capital-goods and noncapital-goods end uses, and we then examine individual end uses.

Capital-Goods and Noncapital-Goods End Uses: Alternative Modes of Market Adjustment

The shares of the major end-use categories as of 1974 are as shown in Table 9.1. As discussed below, in the capital-goods uses, platinum metals are used in production processes where they are dissipated more or less slowly. Recycling is standard practice, and its intensity and speed can be varied. Capital-goods use accounts for more than half of both platinum and palladium use. Firms in these industries inevitably maintain substantial stocks of metals in use, and so can endure disruptions of varying lengths in the supply of these metals.

Our estimates of industry stocks in use, and of dissipation and recovery rates, and our discussions with industry experts indicate that all capital-goods end uses could continue their recent production levels for over a year with little new supplies of metal. It is highly significant that these uses are the most critical in terms of a relative lack of good substitutes.

The capital-goods nature of the more critical uses affects the impacts of disruptions in yet another way. When prices change or are expected to change, using firms not only will adjust their purchases for current use, but also will change their desired levels of stocks. If prices are expected to rise, firms will rapidly build up stocks in anticipation of the increase. And if prices rise, but are expected to fall, firms will wish to lower stocks, which will force prices down. Thus, a country trying to raise prices may be thwarted as the stock-heavy users unburden themselves; the Soviet attempt to raise palladium prices in 1974 illustrates this effect. Suppose a new radical regime in South Africa were to attempt a partial or complete embargo; users, if they expected that prices would soon return to normal, would stop adding to stocks and would sell off some of what would be, at the higher prices, excess stocks. As a result, in the absence of speculative hoarding, price increases would be moderated, and during a partial or selective embargo, producing countries could find sales falling substantially.

TABLE 9.1

Shares of Platinum and Palladium Sales Accounted for by Consuming Industries, 1974 (percent)

Industry	Platinum	Palladium
Capital-goods uses		
Chemical	22.9	18.5
Petroleum	14.8	1.7
Glass	7.9	1.1
Electrical	10.4	44.0
Total	56.0	65.3
Noncapital-goods uses		
Dental and medical	2.7	14.0
Jewelry and decorative	2.4	2.4
Automotive	37.1	15.9
Miscellaneous	1.8	1.4
Total	44.0	33.7

Note: Columns may not add to 100 due to rounding.

Source: Charles River Associates, Policy Implications of Producer Country Supply Restrictions (Cambridge, Mass.: CRA, 1976), p. 211.

Unfortunately, the impacts of speculative hoarding and dishoarding are impossible to predict. While dishoarding in 1974 prevented the Soviet Union from substantially increasing palladium prices, in a future crisis different expectations could well lead to the opposite result. If market participants and speculators feel that a disruption will be protracted, they may well attempt to add to stocks during an emergency, driving prices significantly higher than might otherwise be expected.

Of course, embargoes or price manipulations could be sustained by deliberate producer-country policy. Ultimately, such actions would cause both sharply higher prices for platinum-group metals and efforts to shift to substitutes. A total embargo persisting over a period of several years could cause great economic trauma for the United States. But the large stocks in use and user reactions to actual and anticipated price increases could, under some circumstances, make short-run actions both relatively ineffective and quite costly to their initiators. We now turn to the individual end-use categories.

Catalytic End Uses

Catalysts are substances which can initiate or speed up chemical reactions while remaining chemically unaffected themselves. Platinum is a superior catalyst and is used in pure form and with other platinum-group metals in a variety of applications. Although its use is limited by its cost, platinum is the only known catalyst for some reactions. It can generally be recovered and recycled with small loss. Catalytic uses of platinum now account for over 50 percent of total platinum use. The chemical and petroleum industries use the metals as capital goods, while in the automobile industry, the metals are incorporated in catalytic converters. Substitution possibilities are quite varied and not uniformly well known; a number of government actions could help increase knowledge about alternatives.

Petroleum Refining

Platinum metal catalysts are used in several major petroleum-refining processes, particularly catalytic reforming, the key process in producing unleaded gasoline. The petroleum industry accounts for almost 15 percent of industrial platinum use and 2 percent of palladium use. Technical improvements have already lowered the platinum requirements per barrel of reforming capacity by a factor of four, and further improvements are expected. Losses in recycling are less than 5 percent. Stocks in use tend to be high, and industry sources indicate that even during a total embargo, capacity operation could continue for one to three years.

Research is under way to develop both base-metal and rare-earth catalysts. Industry sources indicate that while platinum metal catalysts last three to five years, the alternatives last only a few hours to a few weeks. Under an extreme supply disruption, the industry could turn to a Mobil Oil Corporation process that reduces platinum use by over 25 percent but alters the mix of refinery output. Petroleum-industry use of platinum depends heavily on the need for unleaded gasoline for automobiles with catalytic converters. This dependence provides some flexibility in the face of a supply interruption; postponement or suspension of Clear Air Act emission standards could reduce the demand for unleaded gasoline, allowing the petroleum industry's demand for platinum to fall, albeit at increased environmental cost.

Chemical Industry

Accounting for about 23 percent of industrial platinum demand and 19 percent of palladium demand, the chemical industry uses the metals as capital goods, primarily as catalysts in producing nitric acid and fertilizer. Other chemical-industry uses include catalytic uses similar to those in petroleum refining.

Unlike most catalytic uses, nitric acid production is a quite dissipative use of platinum, and many of the current purchases are for replacement demand. Because of this, private stocks could carry the industry for about a year, though individual firms could be hurt sooner. If refining and government stocks were used, capacity operation could continue in nitric acid production for three to four years. Substitutes could reduce the demand for, or replace, platinum in nitric acid production. One process (Engelhard's "random pack" method of nitric acid production) reduces platinum use by 60 percent. During a supply emergency, two nonplatinum processes would be available, one of which is now in use by Girdler Corporation; the other was developed by ICI Industries. Industry sources have raised questions about costs, but both processes could be made attractive in the event of a supply disruption. Rare-earth and base-metal catalysts, though possible substitutes, either are not currently available or are grossly inefficient.

Automobile Industry

Beginning with the 1975 model year, 80 to 85 percent of all new U.S. cars were equipped with catalytic converters using platinum-group metals. In 1974 the industry accounted for about

37 percent of industrial platinum demand and 17 percent of industrial palladium demand. Because use of platinum in catalytic converters involves inclusion of the metal in the final product, it is thus unlike the capital-goods uses discussed above. In the absence of new metal supplies, converter production could not continue. However, due to platinum's high value, the salvage rate of platinum from converters has been estimated at between 75 and 90 percent.

The future of catalytic converters for automobiles depends on the development of environmental policy and on technical options for pollution control. Sulfuric acid emissions from converters were at one time a major consideration. There are indications that EPA believes that the acid-emission problem has been exaggerated. Industry has begun substituting away from platinum metals, largely with engine and other modifications, including fuel injection, lean-burn engines, innovative exhaust systems, and stratified charge engines like the Honda CVCC. Thermal reactors and base-metal and stainless steel catalysts offer some promise as alternatives.

All these substitutes for the platinum-based converter would be considered, studied further, and perhaps put to use in the event of a supply disruption. In a sense, however, the question of substitutes is academic since emission standards could be postponed or relaxed in case of a supply disruption in the near future. It is even possible, though probably unlikely that if environmental regulations were relaxed in an emergency, platinum and palladium in use in existing automobiles could be recovered for industrial use. The fact that research on substitutes would be stimulated by a supply restriction and would lead to a long-term decline in platinum demand should help discourage a platinum-supply disruption in the near future.

Noncatalytic End Uses

The two major noncatalytic end uses are in the electrical and glass industries, and are of the capital-goods type. The third end use—in the jewelry, medical, and dental industries—is not of a capital-goods type, but is also less critical.

Electrical Industry

The electrical industry used about 44 percent of the palladium and 10 percent of the platinum consumed by U.S. industry in 1974. The metals are incorporated in circuitry and various electronic components, and can be viewed as industrial capital goods; consumption by such users as the telephone industry could be sustained even with much-reduced flows of new metal.

The Bell System is the largest single consumer of palladium in the United States and accounts for virtually all electrical-industry demand. The metal has been used in mechanical telephone-switching equipment since at least the early twentieth century. Only one palladium-conserving alloy has been found that meets the Bell System's requirements, a silver-palladium alloy of about 60 percent palladium and 40 percent silver. The technology to use it has been available for years, and substitution is well under way.

Currently, most switching systems are hybrids of mechanical and electronic components. Use of mechanical components that contain palladium significantly increases reliability, and complete omission of palladium leads to various problems. It is not clear that the substitution away from palladium-based switching systems can be accelerated readily. The long-run trend is toward electronic switching, but many mechanical switches still are being produced. Of new mechanical switches, 80 percent use the silver-palladium alloy. Given the capital-goods nature of these switches, the superiority of mechanical switches in some applications, and the large capital investment required for a complete changeover to electronic switching, it seems clear that palladium-based switches will continue to be important in the near future.

Industry sources report that the Bell System supplies 40 percent of its own palladium needs with recycled materials. In case of a supply disruption, the Bell System could resort to various measures, such as using more scrap, but might eventually be forced into costly redesign of equipment and use of palladium-free systems that involve a consequent decrease in reliability.

There is widespread use of platinum metals in electronic circuitry, where the main substitutes are gold or gold alloys. In the event of a long-run disruption of platinum metal supplies, it is not completely clear what the impact would be on many of these minor uses. Discussions with industry experts indicate widespread belief that substitution is simply a matter of economics. Nevertheless, the multiplicity of minor uses is such that a more detailed inventory of uses and an investigation of the impact of supply disruptions might be desirable.

Glass Industry

Platinum is used in the glass industry because it has the same expansion characteristics as glass and resists corrosion and heat. In 1974, the glass industry accounted for about 8 percent of platinum use and 1 percent of palladium use. These rates were above normal, reflecting rapid capacity expansion. The metals are used as capital goods and the loss rates are low, between 0.75 and 3.0 percent per year. Industry stocks in use and inventories are large, and production could continue for several years without new supplies. In the event of a long-run disruption, alternative alloys are available, though at higher costs.

Medical, Dental, and Jewelry Uses

In 1974, medical, dental, and jewelry uses accounted for 5 percent of platinum use and 16 percent of palladium use. As industrial production and employment do not depend on the decorative uses, such as jewelry, much of this use category is important in evaluating overall vulnerability to supply disruptions. Industry sources report that palladium in some of these end uses is beginning to be replaced by less expensive stainless steel and other alloys.

Base metals may replace precious metals in dental applications. Industry sources report that research is under way to develop alloys of chromium with cobalt or nickel as potential substitutes for precious metals. Stainless steel may make further inroads into the market. It has also been reported that alloys with precious-metal content significantly lower than substances previously available are being introduced into current use. However, use of nonprecious-metal substitutes for platinum and gold in dentistry often involves fabrication and finishing costs that raise the total cost of using base-metal alloys to approximately the cost of using precious-metal alloys. Substitution could also be hindered by expensive, time-consuming retraining of dental-laboratory technicians.

Once again, however, discussion of such end uses might be somewhat academic. In the face of a supply disruption, dental and especially jewelry uses would probably be curtailed, as a result of either market responses or government action.

Overview: User Alternatives and Responses

The nature of U.S. consumption and the situation of major user industries would tend to moderate the impacts of a short-run, economically motivated commodity action directed solely at the United States. Capital-goods users could postpone purchases, and much higher prices might release substantial metal supplies from jewelry use in Japan.

Examination of individual using industries leads to several conclusions. First, it is clear that, with the exception of the auto and electrical industries, many major users maintain sufficient stocks to permit full production for one to three years after the start of a commodity action. Second, due to the high cost of platinum metals, research is under way in virtually all fields to develop substitutes for these substances. Third, this research has produced substitute materials and processes that are already being phased in, are on the shelf and available for use, or are in an advanced stage of development in most areas.

Of all using industries, the electrical industry would suffer most from a supply disruption by being forced to sacrifice reliability in telephone-switching systems. The automobile industry, though potentially vulnerable due to reliance on catalytic-converter technology, would be unharmed if standards for auto emissions were relaxed. Both the chemical and petroleum-refining industries have substitutes for platinum-group metals available, though it is not clear whether they are fully feasible or economical at this time. The availability of government and dealer holdings of stocks, cutbacks in low-priority uses, and increases in secondary recovery, in addition to use of consumer stocks, could probably sustain even the auto and electric industries for at least two to three years. After that time, major technological alternatives to platinum-using processes, particularly in auto-emission control, would probably be available in at least some uses.

SUPPLY DISRUPTIONS: ECONOMIC IMPACT AND GOVERNMENT POLICIES

The possible impact of platinum/palladium supply problems is much different from that of other critical imported materials, such as chromite and manganese, in which continued production requires a flow of new material, either out of inventories or from imports or other sources. Most of the more critical capital-goods end uses of platinum-group metals could maintain production for some time in the absence of new metal supplies, which implies that the purchases for consumption by using firms in themselves might not exert large upward pressure on prices during an embargo of short duration. In the case of a one-year embargo, the change in prices from this factor alone is estimated to be modest, although this result reflects only the actions of firms who hold platinum as a stock in use. However, this assessment depends on most users' believing the embargo would be temporary, and it also assumes that the government would not threaten controls or other measures that would fuel speculation.

Stock accumulation by other users and by speculators would probably raise prices significantly. In addition, the principal firms' expectations of future prices play a crucial role in the analysis. If further substantial price increases come to be anticipated, firms will have an incentive to hold more stocks in use since the cost of holding them is lowered relative to the anticipated gain. This implies that firms will act to insulate themselves against price increases. The net effects of these factors could cause a substantial increase in prices during an emergency, though the real cost impact need not be large. If buyers' expectations were correct, and embargoes or price hikes were temporary, the resulting purchase deferral and stock liquidations would rule out large impacts.

If a cartel or unilateral supply restriction drastically raised prices and was able to maintain them, sales would be much lower for a time. Once users believed the increase to be enduring, purchases would resume, though at lower levels than would otherwise have occurred; secondary recovery would increase, new substitution begin, and ongoing replacement and economizing accelerate. If prices doubled, the net cost increase to the economy would be $500 million per year, assuming imports remained constant, after the initial period of low purchases. Substitution and technical change could greatly reduce this cost: catalytic converters alone account for over 40 percent of current U.S. industrial use of platinum.

If an embargo lasted for more than two years, many industries could be seriously hurt as their stocks in use would be insufficient to meet requirements. The pressure on prices would be much stronger because the firms would risk incurring large costs from lost output if they did not obtain sufficient metal supplies. The end uses likely to be most severely affected in this situation would be those requiring platinum for catalytic applications.

To guard the U.S. economy against such situations, there are certain government policies that could be implemented. The policies we considered fall into four groups: incentives for the development of domestic resources; tariffs; stockpile programs; and technological incentives and programs.

Incentives for the Development of Domestic Resources

The domestic resource base is so sparse that policies to encourage its development offer little promise. Enormous subsidies of several times the recent prices would be required for production, and even then, the time lags required to bring this production into use would be very long. The risk of lasting disruptions appears too small to justify the extra cost and the creation of the vested-interest and policy-dependent group that would result. The cost of platinum metals has been so high that most stocks are heavily recycled. There appears to be no need for government incentives to increase secondary recovery.

Tariff Policies

For many years the United States has had tariffs on imported platinum-group metals, which vary according to the form in which the platinum is imported and the origins of the metal. The exact reasons for the tariffs' existence are obscure. Possibly tariffs were originally imposed to encourage the development of domestic supplies or to encourage industry to develop alternatives to platinum. Whatever the reasons, it appears that tariff policy will have little or no direct effect on either the development of domestic resources or in reducing purchases or consumption for some time. Domestic resources are too sparse for reasonable tariffs to yield worthwhile increases in production. If the government announced an impending tariff hike, users would quickly attempt to build up stocks, and thereafter their purchases would be affected only moderately and in the long run. In the critical capital-goods end uses, it is unlikely that plausible tariff levels would change decisions on capacity expansion, and therefore purchases of metal would continue for some time.

Stockpile Policies

For years the United States had maintained sizable stockpiles of platinum and palladium as part of its strategic commodities program. Currently, the stocks held by GSA are 452,000 troy ounces of platinum and 1,255,000 troy ounces of palladium. Prior to 1976, surplus stocks (in excess of strategic goals) were sufficient to satisfy domestic consumption for one year for the more critical end uses; however, strategic goals were raised in October 1976, creating deficits for both materials approximately equal to a year's consumption.

Our analysis of the efficient stockpile releases for varying embargo durations, likelihoods, and price effects indicates that in the absence of speculative hoarding, the current levels of government stockpiles—if they were available for economic emergencies—would be at least adequate to guard against the possibility of supply disruptions, or cartel actions, even under probabilities of disruptions that were as high as 20 percent per year. The probabilities for disruptions in the palladium market are higher than those for platinum, judging from the attempt by the Soviet Union to nearly double palladium prices during 1975-76. Yet the stock liquidations that defeated that attempt indicate that using firms have substantial amounts of protection on their own.

Because the effects of speculative hoarding are impossible to predict with any precision, it is difficult to reach conclusions about the appropriate magnitude of public contingency stocks. If speculative hoarding were not a factor, the capital-goods nature of platinum and palladium would lead to a conclusion that significantly smaller stocks of these are more appropriate than, say, for chromium and manganese. However, speculative hoarding would almost certainly be an important factor during an emergency, and an economic contingency stockpile equaling as much as one year's consumption (equal to the excess stockpile before October 1976) might well be warranted. Direct measures, such as price controls, might be attempted, though these could have a number of undesirable side effects.

Technological Incentives and Programs

An adequate stockpiling program could moderate the impact of a disruption lasting as long as several years, and high-cost alternatives do exist for many platinum and palladium users, should a disruption last longer. Furthermore, the availability of stocks in an emergency would buy time during which a crash program for substitution could be launched. As a result, while some government funding of basic research and development may be justified because of the public-good aspect of such investment, expenditure of substantial public monies on a crash program to substitute away from platinum and palladium is probably not warranted at this time. And such public funding could reduce private sector efforts.

While many of the alternatives to platinum/palladium-based processes are well known, information on some is closely held. One implication of our research is that government might increase the diffusion of information about the processes, while remaining sensitive to problems of patent policy and trade secrets.

In the automobile industry a number of technical alternatives to catalytic converters are on the market or are being developed. Here, as in other applications, the feasibility and economy of base-metal catalysts remain uncertain. Because of the multiplicity of minor electrical uses, a detailed inventory of uses and the possible impact of disruptions could be desirable.

Catalytic Applications: Petroleum and Chemical Uses

The situation among the catalytic uses of platinum-group metals in the petroleum and chemical industries is relatively unclear because of the enormous range of catalytic processes, each of which has specific requirements. Much research is being devoted to the development of catalytic processes that do not rely on noble metals.

Because the returns on such research are potentially great, information on the current state of technology is fairly closely held. Nevertheless, the information we obtained through a series of discussions with catalyst experts was remarkably consistent. Given current technology, the noble metals are far superior to any alternative materials for use in catalysts; although current types of alternative catalysts could be used in the event of an emergency, using them would be costly, would involve substantial alteration of existing processes, would take a number of years to implement, and would require much capital investment.

Some experts believe that catalyst technology based on rare earths represents a feasible long-run alternative to platinum metals. Rare earths have the advantage that they are available in abundant supply, are likely to be cheaper than platinum-based catalysts, and would be compatible with existing technology. The feasibility of rapid supply expansion, however, is not certain. Private industry sources expressed some skepticism about this alternative because of the expected rapid deterioration of rare-earth catalysts, although at this stage, not enough is known to make an accurate assessment. The development of these catalysts is expected to take $5 million over a 25-year period. It seems likely that pooling and verifying of information about catalytic alternatives would be worthwhile.

While all indications are that industry could protect itself against short-term disruptions, the catalytic uses appear to be the most vulnerable because they are so crucial, and because alternative technologies do not seem sufficiently well developed to provide protection to industry against long-term disruptions.

General Government Policies toward Technology

In view of the foregoing discussion of the technological alternatives to platinum-using processes, a sensible government policy would be to develop general contingency plans to cover the transition to alternative technologies, where they exist, in the event of a long-run disruption of supplies. While our analysis does not indicate the direct need for any active, current intervention on the part of the government to protect the platinum-using industries, in the short run, there should be contingency plans to cover the transition to alternative technologies and, possibly, to control nonessential uses of platinum.

The Bureau of Mines plans to spend $200,000 annually over the next 25 years for research on rare-earth catalysts. The present discounted value of this research is only $2 million, assuming an interest rate of 10 percent. If this is compared with the present discounted value of the cost of any supply disruption that permanently raises the cost of platinum metals even slightly, the discounted costs of the expenditures on research and development will be less than the discounted costs of the disruption. The appropriate comparison to make, of course, is between discounted benefits of the research, as measured by the change in costs of the new technology over the old technology, and the discounted costs of the expenditure on research. Such a comparison is not possible, given our current state of knowledge. Nevertheless, in light of the changes in the cost of using platinum metals due to even short-run supply disruptions, the potential benefits of these research and development expenditures might well exceed their cost.

NOTE

1. U.S. Department of the Interior, Critical Materials: Commodity Action Analyses, Aluminum, Chromium, Platinum (Washington, D.C.: Government Printing Office, March 1975), p. IV-42. Since our study was completed, additional information on the stillwater deposit has become available which indicates that at the higher prices of 1979-80 the deposit might be commercially feasible. If so, production from the source might ultimately provide about 20 percent of U.S. platinum consumption and nearly all of U.S. palladium consumption.

PART V

COPPER: A CARTEL POSING LITTLE THREAT

While copper is a large item in U.S. and world trade, import dependence is low, and domestic resources are large and usable at reasonable costs. Furthermore, reasonably efficient substitutes are available for most uses. While the current major suppliers of copper to world markets are primarily developing nations, Canada and the United States are leading producers. These factors, combined with the demonstrated eagerness of exporting nations to expand production, indicate that the United States faces little threat.

In this sense, and in view of the many other dimensions that determine threats, the copper market appears more typical of world material markets than most of the other markets studied in this book. For example, substitution away from tin in the long run moderates the potential power of the International Tin Agreement. The tungsten market is not highly concentrated, substitutes are numerous, and the market is highly unstable and difficult to control. The nascent Tungsten Producers Association would seem to pose little threat. Finally, while a number of countries have joined the Association of Iron Ore Exporting Countries, they command a small share of the world market, and the United States and other industrialized countries are the major producers. Thus, our central modest policy recommendation for copper—that some stocks, not necessarily all public ones, be held—may have rather wide applicability.

10
COPPER: IMPACT AND POLICY ANALYSIS

As discussed in Chapter 3, the United States is currently largely self-sufficient in copper in spite of CIPEC, the ongoing producer association that occupies a prominent position in the world copper market. Furthermore, the costs of significantly increasing domestic production are not greatly above world price levels. Substitutes are available for many end uses at moderately elevated costs, and CIPEC appears to have neither the power nor the motivation to engage in severe restrictions.

The CIPEC countries have a large share of total non-Communist-world exports, a fact which has led some researchers to conclude that CIPEC represents a viable threat to the United States and other consuming nations. This conclusion is unfounded since the export market is not the relevant market for analysis. Instead, at the least, the entire non-Communist market must be considered. Were CIPEC sharply to restrict its exports, other countries could increase their production and exports and reduce consumption. Considering all of the consumption and production alternatives that exist to CIPEC copper, over reasonable periods of time, output cutbacks would significantly reduce CIPEC revenues.* Simulation of CRA's copper market model and analysis of producer-country expansion plans support the conclusion that drastic price escalation or embargoes are most unlikely under current market conditions. No one CIPEC country has the market

---

*In technical terms, we estimate the one-year price elasticity of CIPEC demand is -1.00; the five-year elasticity is -2.98 and the 30-year elasticity is -5.31.

share to impose a damaging embargo—Chile, the largest, has a share of only about 13 percent. This market structure distinguishes the copper market from the cobalt, platinum, chrome, and manganese markets, where the loss of production from one major country could have dramatic effects. Similarly, no CIPEC country could act as a cartel stabilizer by holding large amounts of excess capacity, as Saudi Arabia has done in the case of the oil market.

Higher domestic production costs, caused in part by lower ore grades and environmental control costs, will probably lead to a steady increase in the import share in U.S. consumption over the next several decades. If foreign production becomes increasingly concentrated in the CIPEC countries, this would lead to a situation in which security of supply problems become much more serious than at present. This suggests that government policies to encourage technological adaptation designed to slow this erosion in self-sufficiency may yield significant security benefits in the future.

We appraised the economically efficient stockpile, tariff, and subsidy policies under extreme assumptions of highly unlikely complete embargoes. We also appraised the general usefulness of encouraging solution mining, a technique to increase production from low-quality domestic ores. In the remainder of this chapter, we describe the likely range of future U.S. import dependence, briefly summarize the viability of substitutes for copper, and present policy conclusions.

It should be noted that the fundamental difference between copper and the other strategic materials studied in this book leads to quite different policy problems. In the other materials, the United States is highly dependent on imports from insecure supply sources, and the policy issues revolve about ways of reducing the expected costs of this insecurity. In copper, we are currently largely self-sufficient. The major policy issues arise as a result of the problems posed by potentially excessive low-cost foreign suppliers that could lead to lower U.S. production, unemployment in copper-producing areas, and potential security of supply problems 10 or 20 years in the future. Protective actions (such as tariffs) may well be justified to alleviate unemployment problems in the present; analysis of such policies was beyond the scope of this study.

LIKELY U.S. IMPORT DEPENDENCE AND THE THREAT OF RESTRICTIONS

The United States would face substantial threats only if import dependence rose. In 1974 net copper imports were only 11 percent of consumption. Rising imports would indicate low foreign copper prices relative to domestic prices and costs. U.S. expansion plans indicate about a 4 percent growth rate in mine capacity. Smelting and refining capacity are likely to grow more slowly than consumption in the long run, in part because of environmental-policy restrictions. As these restrictions have become tighter, and as mining has shifted toward increasingly lower-grade ores, costs have risen.

If CIPEC output grows at the 9 percent rate consistent with capacity expansion plans between 1975 and 1980, the United States will become more dependent on imports; foreign costs tend to be lower than domestic costs. Therefore, import dependence depends on CIPEC policies. Growth in foreign production will keep real prices low enough to preclude large additions to U.S. capacity, and by 1990 import dependence is projected to be as high as 47 percent of domestic consumption, with most of the increase coming after 1980. Such import dependence could subject the United States to risks of short-run interruptions in copper supplies. This possible future change in import dependence is, in part, the result of current CIPEC expansion.

We believe that the 9 percent CIPEC growth to 1990 and the resulting 47 percent import dependence are too high. First, such growth keeps the real copper price at about $0.72 per pound between 1981 and 1990 (in 1974 dollars). But because costs of copper in typical producing areas appear to be about $0.80 per pound, production costs would dictate slower output growth. In addition, if CIPEC is at all effective, we would expect it to constrain production to raise revenues. Simulations of the market indicate that slower (6 percent) growth after 1980 would yield roughly maximum copper revenues. This path keeps copper prices at about $0.87 per pound, in 1974 dollars, which appears sufficiently high to encourage some additions to U.S. capacity, and results in 1990 import dependence of about 32 percent.

There is a certain dilemma inherent in these results. To the extent that CIPEC becomes an effective cartel and maximizes revenues, U.S. import dependence will be lowered. If CIPEC members compete vigorously among themselves and expand capacity, U.S. import dependence may be raised more substantially, as will exposure to future threats.

THE SUBSTITUTES FOR COPPER AND THE RISKS FROM SUPPLY INTERRUPTION

Based on our econometric and qualitative investigations of major end uses, we conclude that substitutes and replacements for copper are, in many uses, substantially more economic and desirable than those for several of the other materials studied, including chromite, manganese, and the platinum group. Copper is valued for its combination of strength, ease of forming, corrosion resistance, and thermal and electric conductivity. Copper's low electrical resistance is its economically most important characteristic; about half of all copper consumption is in applications relying on electrical conductivity. Aluminum, the major substitute in these end uses, has about the same strength and hardness as copper and lower conductivity. Substitutes are available in all major end uses: aluminum can be substituted where conductivity is important; where conductivity is not important, but corrosion resistance is, stainless steel, other alloys, or plastics might be considered.

The substitutes are not necessarily similarly cost effective or instantly available. For example, switching to aluminum tubing in air conditioners or in auto radiators would require significant investments, lead times, and extra costs, a fact which is verified by our econometric estimates of price elasticities of demand. We found that in the short run, copper use was quite insensitive to increases in its price. Estimated elasticities ranged from -.003 for consumer products, such as appliances, to -0.1 for transportation. In the long run, the elasticities were much larger, from -0.1 in consumer products to -0.7 in transportation. These results indicate that while substitutes are not perfect, they are much more available in the long run than in the short run. For quite substantial price increases, of the size that could result from a supply disruption, the degree of substitutability is almost surely higher than these estimates indicate. The estimates are based on normal responses to typical price increases, while major escalations would provoke larger responses. This finding is consistent with those of the National Materials Advisory Board.

The degree of substitution possibilities for copper in the long run indicates that holding some stocks, if disruptions appear likely, might be beneficial. Such action would minimize the problems resulting from short-run inflexibility. Nevertheless, the problem is less severe than for the other materials examined in this book.

STOCKPILE, TARIFF, AND SUBSIDY POLICIES TO MITIGATE THE IMPACT OF SUPPLY DISRUPTIONS

Our analysis of the market indicates that substantial disruptions in the copper market are unlikely under current conditions. We evaluated policies to guard against a total embargo, not because such an embargo is a realistic possibility, but because we wished to explore the worst possible circumstances.

We evaluated the policies for three scenarios of import dependence: high dependence (50 percent), medium dependence (25 percent), and low dependence (5 percent). We assumed that the probability of a total embargo was 20 percent in any given year (that is, on the average of once every five years). Note also that this means the probability of experiencing at least one embargo in five years is .66, and one in ten years is .89. Further, we assumed that once an embargo started, it had a continuation probability of 80 percent each year after the initial year (that is, 80 percent probability in year two; 64 percent probability in year three). The probabilities both of an embargo's starting and of its continuing are assumed to be much higher than in fact are likely to be reasonable under almost any imaginable circumstance.

Impacts and policies were evaluated using the dynamic policy-evaluation program described in the Appendix. Economic losses consist of increased domestic production costs and losses to users from shifts to less cost-effective substitutes and from foregone consumption.[1] Policies are chosen to achieve the maximum net reduction in this loss, taking into account policy costs, and the results indicate those policies which are, on average, economically most beneficial.

The results for the three sets of calculations are presented in Table 10.1. The long-run autarky situation shown in each case corresponds to the estimated prices and quantities that would exist if the United States did not import any copper. The high-import-dependence case has a high long-run autarky price since the assumed demand and supply correspond to a situation in which the United States is very import dependent.

In the high-dependence case, the optimal tariff and subsidy are each about 25 percent of the initial price; they raise the price to about half of the autarky price. Hence, even in a high-import-dependence scenario, the optimal policy measures are not extreme, though the sums involved are substantial. The redistributions involved are significant: at the original consumption level, the annual subsidy to primary producers is about $343 million (in 1974 dollars), and the subsidy payments will rise as production rises. The optimal stockpile in the higher-import-dependence case is equal to about one year's consumption of copper at the initial price of $0.88 per pound (in 1974 dollars). Hence, we see that in the high-import-dependence scenario, most of the protection from an embargo comes from the copper stockpile, not the tariff. This conclusion is reinforced by considering the medium-import-dependence scenario.

In the medium-import-dependence scenario, the roles of the tariff and stockpile are reversed. In this case the tariff of $0.15 increases the price of copper to 80 percent of the long-run autarky

TABLE 10.1

Optimal Tariff/Stockpile Results for Three Copper Scenarios

Scenario	Price (1974 dollars per pound)	Primary Supply	Scrap Supply (tens of thousands of short tons)	Total Demand (tens of thousands of short tons)	Imports	Optimum Policies* (1974 dollars)	
High import dependence							
Original	.88	78	286	762	398	$0.21	tariff
Long-run autarky	2.03	140	84	224	0	$0.22	subsidy
						750.6	stockpile (tens of thousands of short tons)
Medium import dependence							
Original	.88	178	183	488	127	$0.15	tariff
Long-run autarky	1.29	229	137	366	0	$0.14	subsidy
						48.9	stockpile (tens of thousands of short tons)
Low import dependence							
Original	.88	218	142	379	19	$0.03	tariff
Long-run autarky	.94	228	137	365	0	$0.02	subsidy
						0	stockpile

*The optimal policies are calculated under the assumption that the inception probability is 20 percent for a complete cutoff of imports, and that the continuation probability is 80 percent. These probabilities are unreasonably high, and the option tariffs and stockpiles should therefore be regarded as maximum estimates.

Source: Charles River Associates, Policy Implications of Producer Country Supply Restrictions: Overview and Summary (Cambridge, Mass.: CRA, 1976), p. 230.

price, while the stockpile is equal to only 10 percent of the initial consumption rate.

In both the high- and medium-import-dependence cases, the tariff increases the copper price to between $1.00 and $1.10 per pound (in 1974 dollars). The size of the stockpile, however, differs widely. Since the medium-dependence scenario is probably closest to the probable market situation in the early 1980s, we conclude that, at the very most, a small tariff and stockpile are warranted at present for strategic reasons. If imports rise over time, the optimal stockpile and tariff levels would also rise.

Government Aid for Technological Innovation

Embargo threats depend on import dependence. The expected future increase in import dependence in part is based on rising costs of domestic production. Innovations in production could offset this tendency and reduce imports. Two possibilities are solution mining and ocean mining.

Solution Mining

Solution mining is the term applied to a group of techniques based on leaching the copper in place (separating copper from the ore with an aqueous solution). Solution mining applies the same principles as the normal in situ leaching to very deep deposits. Kennecott evaluated a program to extract a low-grade copper deposit, 5,000 feet underground, by solution mining. The leaching solution would be pumped down to the deposit and returned, and at the top of the return cycle the copper would be recovered, and the solution would then be returned to the pumping cycle. Because ore is not moved in solution mining, ore grade is much less important in determining relative mining costs than is the case for traditional mining techniques. As a result, solution mining may restore the U.S. competitive advantage in copper mining.

Very little of the specifics of solution mining is publicly known. Solution mining is apparently virtually pollution free and potentially much less costly than conventional deep-underground mining. If the process proves economic, it could presumably open up large new copper resources in the United States for recovery. However, data indicating the quantity of the resources that could be exploited profitably using the technique are not available.

If solution mining proves economic, the long-run dependence of the United States on imported copper would be reduced since domestic production would rise. While solution mining will increase the long-run independence of the United States from imported copper, it would presumably be of little short-run help in the event of an embargo in the near future, because of the long lead time necessary to bring a new mine into production.

Ocean Mining

Ocean mining will have little impact either on the future copper market or the results presented in this chapter. Ocean mining probably will not be economic until 1990, and not in widespread use until 1995. The operating scale envisioned for ocean-mining plants indicates that each plant would recover between 10,000 and 30,000 tons of copper per year. That is, one project, at a maximum production of 30,000 tons per year, would only represent 0.45 percent of world copper demand in 1990. In terms of production capacity, an ocean-mining plant is equivalent to a moderate-sized conventional copper mine, and hence copper-market risks alone probably cannot justify government assistance.

CONCLUSIONS: CONTINGENCY POLICIES IN THE COPPER MARKET

The United States is currently nearly self-sufficient in copper. Unless some additional improvements in U.S. mining technology are forthcoming, imports are likely to grow in the future. The question is not one of total copper availability, as the United States has substantial reserves; rather, the issue is the price at which additional resources will be mined. Foreign copper will be more competitive domestically in the future unless cost-reducing improvements in domestic copper mining and processing are found. Environmental-control and -protection costs, which are of course affected by government policy, are significant in this regard.

The calculations of the economically efficient tariff-stockpile policy under the different scenarios give an indication of the likely costs imposed by embargoes. In the moderate-import-dependence case, the current efficient tariff is $0.15 per pound (in 1974 dollars), and the current efficient stockpile is 489,000 short tons.

The increasing dependence of the United States on foreign copper may well lead to security of supply problems in the future. This suggests that policy makers should monitor the copper market and that we should be prepared to consider protective actions if import shares begin to increase substantially.

Solution mining, and to a much lesser extent ocean mining, offer substantial promise as technologies which can prevent the decline of the U.S. copper industry. Government encouragement of these promising technologies may well yield substantial contingency benefits in the future.

NOTE

1. Since we are dealing with total embargoes, there are no increased payments to foreign producers. The method is fully described in Charles River Associates, A Framework for Analyzing Commodity Supply Restrictions (Cambridge, Mass.: CRA, August 1976), chap. 6.

APPENDIX: MODELS OF OPTIMAL POLICY

This appendix describes the optimal policy model or program we used in the text to derive optimal stockpiles under various scenarios. We used several variants of the model. The most complete version takes into account two particularly important effects of stock holding and releasing: price-deterrence effects, whereby stock releases during a disruption result in lower prices on the world market; and probability-deterrence effects, whereby larger stockpiles held prior to a disruption decrease the likelihood that a disruption will in fact occur.

As described in the text, we designed the model to estimate the expected economic costs of crises, as well as the expected benefits of policies such as stockpiles, tariffs, and quotas. The components of cost (and hence benefits, which are, after all, reductions in cost), include increased payments to foreign producers; increased costs because of resorting to inefficient domestic supplies; and costs of users' restricting consumption or resorting to less desirable alternatives.

In addition, the program allows for general effects of U.S. policies on the likelihood and cost of a crisis—the probability-deterrence effects and price-deterrence effects described above. These deterrence effects create the opportunity for additional stockpiling to benefit economically the United States; in such cases, private stockpilers will generally have no financial incentive to carry as much stocks as would be optimal from the point of view of the country as a whole.

The probability-deterrence effect would operate when larger U.S. stockpiles reduce the probability of a disruption's occurring or, once it has begun, reduce the probability of its continuing. This effect is most likely to occur where the disruption is based on attempts by producing countries to increase their economic gain from exports, for example, by cartel formation. It is less likely to be important for an embargo based on larger political objectives. Application of the policy models indicates that the probability-deterrence effect can justify fairly large nonmilitary contingency stockpiles for the United States in certain cases.

Price-deterrence effects operate when the United States remains significantly dependent on imports during disruptions, even after the efficient size of the stockpile has been accumulated. Decreased prices on the world market, as the result of U.S. stock releases during the disruption, then benefit all U.S. importers (as well as foreign users). This effect is most important in markets such as chromium and manganese, for which this country is completely import dependent in normal times, and for which the price elasticities of U.S. demand and supply are very low. Price-deterrence effects remain a relevant consideration for most disruption scenarios, as long as the United States has access to imports during the disruptions.

One of the most important implications of deterrence effects is that they create a divergence between the stockpile size that is most efficient for the United States as a whole, and the size of the stockpile that a competitive profit-maximizing private sector would accumulate. As illustrated in the following discussion, socially optimal stockpiles can be much larger than private optimal stockpiles, suggesting that there may frequently be an important role for government stockpiling. In a similar fashion, deterrence effects can provide a strong rationale for coordination of stockpiling policy among all consuming nations.

We do not address the difficult issues of how government stockpiling should be implemented. There could be a mechanism for using part of the U.S. strategic stockpile. Alternatively, there could be subsidization of private stockpiling or some other decentralized option.

The illustrative calculations of optimal U.S. stockpiles presented below are for the total of private and public stockpiles available for nonmilitary contingencies, assuming that any public stockpiles are efficiently managed. This is the most clear-cut simplifying assumption to make in initially assessing the benefits of various total stockpile sizes for the United States.

The remainder of this appendix reviews the implications of this generalized policy model for stockpiling policy in the markets considered in the body of the text. Since we have applied the most general of the policy models extensively to the chromite market, we have used estimates for that market to illustrate, in some detail, the nature of the model and the types of information that it requires and provides. As we discuss each of the other markets, we present features of the policy model that are particularly important for those applications.

The policy model can be used for such purposes as evaluating the benefits from technological adaptations that lessen dependence on foreign suppliers of raw materials. However, in this appendix, we concentrate on stockpiling recommendations, since we have concluded that stockpiling is by far the most powerful policy tool available to the United States in most of the markets studied. Technological adaptations and other policy options must be more efficient (or desirable on other grounds) than additional stockpiling if they are to be recommended, and usually there is at least the question of coordinating stockpiling policy with whatever other policy tools might be selected.

In the remainder of this appendix, we first describe the general nature of the policy model. We then illustrate in detail its application to an important disruption scenario in the chromite market. Finally, we discuss the rationales for the optimal stockpiling levels presented in this study in general terms, with particular attention to the relative importance of the assumptions underlying the optimal policy model.

NATURE OF THE GENERALIZED CRA POLICY MODEL

In this section, we discuss the structure of the generalized policy model by referring to a series of graphs and by an explanation of the inputs and outputs of the model. The policy model itself is mathematical while the graphs are only heuristic, not a complete representation.[1]

Underlying Market Model

The way in which market characteristics are incorporated in the policy model is illustrated in Figure A.1, which shows the supply and demand situation in the U.S. market for a raw material.[2] In the case illustrated, normal U.S. consumption is much greater than normal U.S. production, and the difference is normally satisfied by imports.

The market disruption of concern occurs abroad, raising the world market price severalfold, from the normal market level to the illustrated disrupted market level. The extent of this price increase is developed by prior analysis, often utilizing a detailed econometric world market model. The world market price may change as the disruption continues, particularly as a result of U.S. and foreign stock releases; for simplicity, such stock releases are considered separately below, and not in Figure A.1. The inception and continuation of a disruption are uncertain events with specified probabilities.

U.S. demand and supply are likely to be fairly unresponsive to price in the short run, as illustrated by the more nearly vertical short-run curves in Figure A.1. Thus, at the beginning of the disruption, imports drop only from the normal quantity indicated by the lowest horizontal bracket to the quantity indicated by the somewhat smaller bracket just above it. In the long run, the domestic demand and supply curves for the short run shift up the respective

FIGURE A.1

Generic U.S. Market Model Underlying the Policy Model

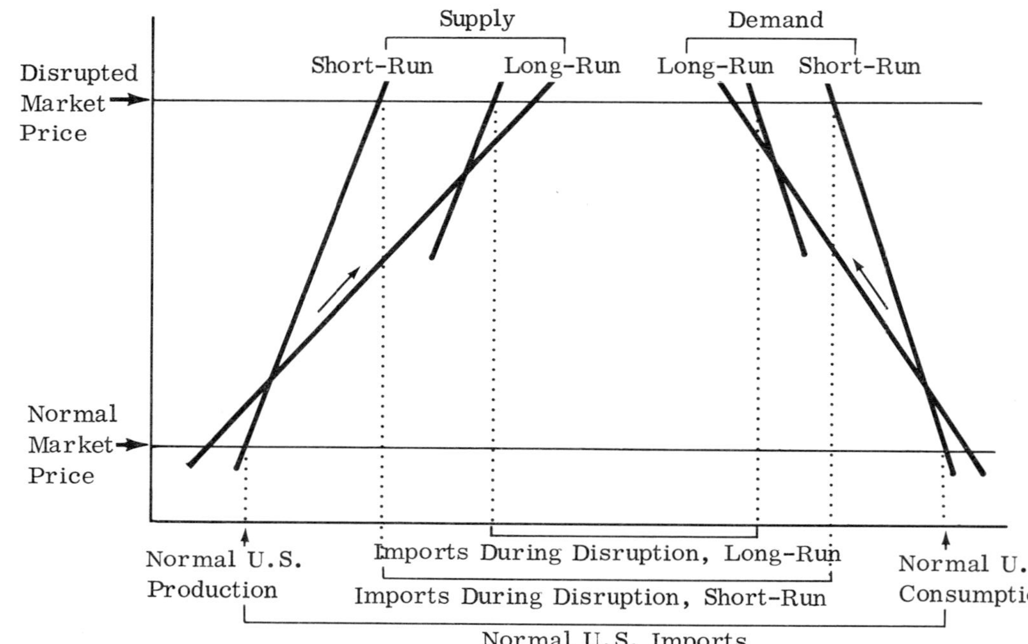

Source: Constructed by the authors.

long-run curves, as indicated by the arrows in Figure A.1, and imports decrease to the quantity indicated by the smallest horizontal bracket.

Types of Losses

Most of the major outputs of the policy model are indicated by Figure A.2, which disregards, for simplicity's sake, the dynamic distinction between short run and long run and the effect of stock releases on the disrupted world price, but does recognize the effect of releases from U.S. inventories and stockpiles on U.S. losses. The economic cost of the foreign supply disruption can be represented geometrically by various areas between the disrupted market price and the normal market price on the vertical axis. These areas are dollar values; they have a vertical price dimension and a (horizontal) quantity dimension.

The reduction in U.S. consumption during a market disruption is indicated on the right side of Figure A.2. The cost of conservation by domestic consumers is represented by the triangle under the demand curve on the right. This triangle conceptually includes the cost to U.S. industry of utilizing substitute materials in production and the cost to ultimate consumers of switching to other finished goods.* The entire rectangle to its left represents the extra payment by domestic consumers for their remaining consumption during the disruption. The parties gaining at the expense of U.S. consumers are indicated by the subdivisions of this rectangle, which we will describe from left to right.

U.S. suppliers will, in many cases, increase their production in response to the higher price during the market disruption. Part of the additional revenue they earn is expended for added costs of production, represented by the triangle under the supply curve in Figure A.2, but the remainder is added profits.† In the standard

---

*The consumers' surplus loss measured by the area of the triangle assumes other input prices are constant during the disruption. This assumption is reasonable, but there may be exceptions. For example, the total consumers' surplus loss in the stainless steel market from a simultaneous crisis in both the chromite and nickel markets cannot be reliably calculated as the sum of the consumers' surplus loss calculated separately for each crisis.

†Most analysts would agree that it is desirable for domestic suppliers to produce at the high point on their supply curve, which corresponds to the disrupted world market price. It is very difficult to design practical policies that reduce domestic producers'

FIGURE A.2

Graphical Representation of Loss Components Calculated by the Policy Model

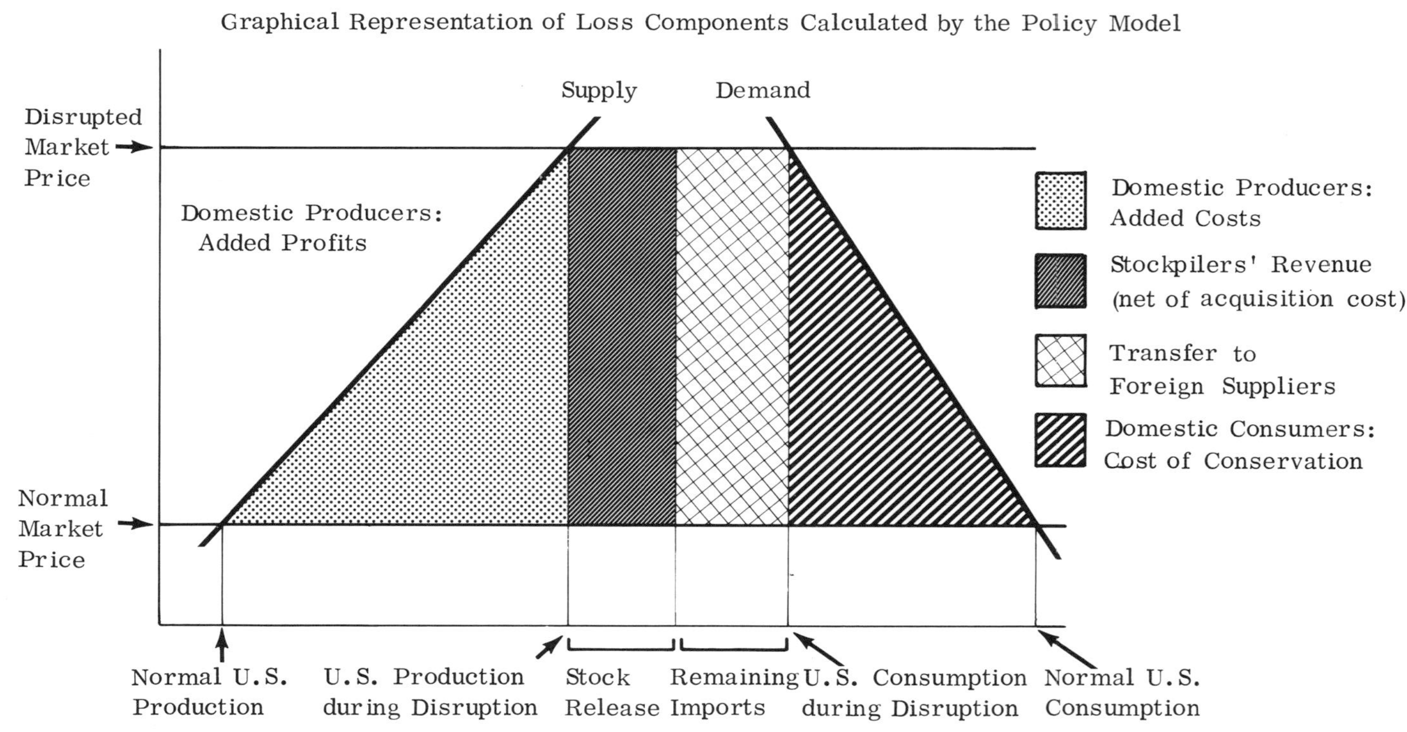

Source: Constructed by the authors.

economic treatment of losses inflicted on the United States by an increase in the price of imports, the transfer of income from domestic consumers to domestic producers is considered neutral since both groups are members of society. Thus, the net loss to the nation is given by the shaded trapezoid bounded by the supply and demand curves and by the normal and disrupted market prices. However, different evaluations of shifts in the distribution of income could be readily incorporated in the analysis.

U.S. consumption that is not satisfied by domestic production must be met from U.S. stock releases or by imports from foreign suppliers. Stockpilers' revenues, net of acquisition costs, are represented by the small rectangle in the middle of Figure A.2, which is so labeled.* Out of their revenues, stockpilers must pay the interest charges on their acquisition costs and the cost for other resources required to operate and maintain the stockpile (for example, land, buildings, security, and supervision). Averaging out years in which prices are normal, and those in which releases can be made at high prices, stockpilers may show either an overall profit or an overall loss. It is impossible to determine this from the information in Figure A.2, though the policy model does provide this calculation.

The "remaining imports" indicated on the horizontal axis of Figure A.2 must be purchased during the disruption at the much higher disrupted market price. The "transfer to foreign suppliers," beyond what this quantity of imports would cost in normal times, is the last of the major categories of economic losses that are indicated in the figure.

---

windfall profits but leave unimpaired the high price incentive inducing this desirable expansion of production. For purposes of discussion here, we will assume that the price system is allowed to operate unimpeded during disruptions. If a government policy such as the multi-tiered-pricing system for domestic crude oil is imposed, then the magnitudes of losses in various categories will change.

*Acquisition costs, assumed to be at the normal market price, are represented by the smaller rectangle just below "stockpilers' revenue (net of acquisition cost)"; stockpilers' gross revenue is naturally the amount of the "stock release" times the "disrupted market price." It is convenient here to speak of stockpilers as though they were separate economic entities with their own accounting records, though of course in practice domestic consuming and producing firms often do most of the stockpiling.

Because the actual policy model is dynamic, unlike the static representation in Figure A.2, it provides two further types of economic loss estimates. Figure A.3 graphically illustrates predisruption costs incurred by U.S. producers and consumers because they recognize the probability that a disruption will occur. If domestic suppliers thought there was no possibility of a major supply disruption, then they would produce the quantity on the long-run supply curve at the normal market price; their plant, equipment, and production practices would be exactly appropriate to that level of production because they would not allow for the possibility that they would want to produce more on short notice. By definition, producers always operate on the short-run supply curve, so this precise coordination of actual production levels and producers' capital stocks would be represented graphically by positioning the short-run supply curve so that it intersects the long-run supply curve at the normal market price.

But, of course, in a real world where future prices may be normal or may occasionally be much higher during world market disruptions, selection of plant, equipment, and production practices (determining the position of the domestic short-run supply curve) must be a compromise. The nature and implications of that compromise are suggested by the position of the short-run supply curve in Figure A.3. Because a larger capital stock shifts the short-run supply curve to the right from its most efficient position at the normal market price, it becomes efficient to produce more when there is a high or "disrupted" market price. However, the larger capital stock usually also implies that U.S. production is greater at normal prices than would otherwise be the case, and that losses are incurred that can be measured by the crosshatched area between the long-run and short-run supply curves in Figure A.3.

The situation on the demand side of the domestic market is symmetrical with that just described for the supply side, so we will not explicate it in detail. By maintaining the capability to use substitutes or forego consumption altogether when prices are high, domestic consumers incur costs during periods of normal prices that can be measured by the crosshatched area between the demand curves.* The losses indicated by the crosshatched areas on both

---

*The policy model employed in this volume uses linear approximations to market supply and demand curves, as illustrated in Figure A.3. For reasons suggested in Figure A.2, it is much more important that the short-run curves be accurate during periods of high disrupted market prices than during periods of normal prices:

FIGURE A.3

Predisruption Costs Incurred by U.S. Producers and Consumers for Capacity Adjustments

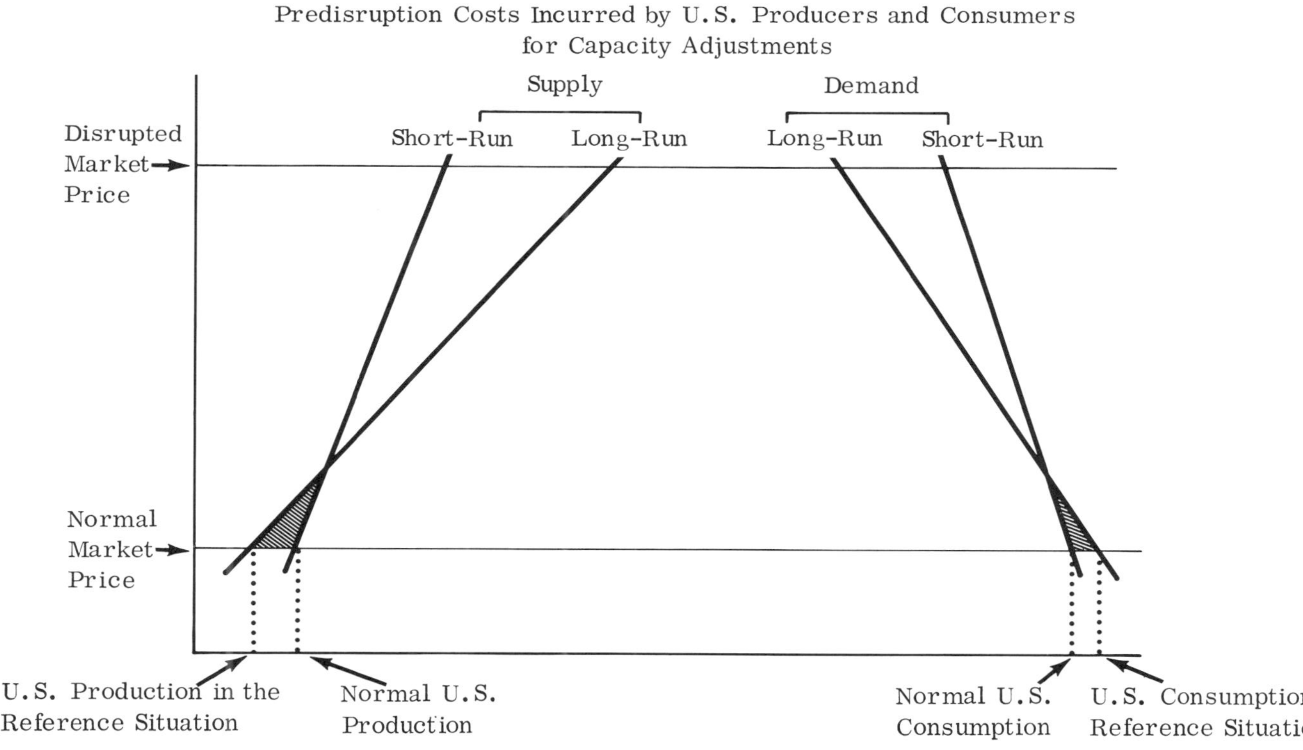

Source: Constructed by the authors.

the demand and supply sides in Figure A.3 are analogous to the cost of maintaining a stockpile, in that they are incurred prior to disruptions. None of these costs would be incurred if there were no threat that a disruption would raise the market price.

The second type of dynamic economic loss that is not represented in Figure A.2 is indirectly suggested by the arrows in Figure A.1. These arrows indicate movement in the short-run supply and demand curves during disruptions. If capacity is expanded very gradually, capital can be acquired and installed at normal costs, which are indicated by the long-run supply and demand curves. However, after a serious disruption begins suddenly, it is usually efficient to expand capacity rapidly, even if extra costs are incurred. The policy model provides rough estimates of these extra costs for accelerated capacity adjustment.*

Uncertainty Recognized

The policy models explicitly recognize the uncertainty typically associated with the occurrence of the most severe market disruptions, although in a particular way. At times when there is no disruption, it is assumed that there is a specified inception probability, indicating that the disruption will begin during the next month (or year).† Thus the possibility of both very short and very long disruptions is recognized.

---

the effect on estimated losses is much greater during market disruption than during normal times. Calculating the cost of predisruption capacity adjustments by measuring loss areas such as those crosshatched in Figure A.3 is intended to provide only a very rough estimate, which should be appropriate within an order of magnitude. Other categories of economic losses are generally much larger, so, typically, the first one or two significant figures of the total loss are not even affected by such a predisruption cost estimate.

*As was the case for predisruption losses due to adjusted capacity positions, the losses in this category are typically quite small relative to other types of economic losses, so a very rough estimate is adequate.

†The most advanced form of the model is formulated mathematically in terms of continuous time. Probabilities can be adjusted so that the choice of time unit makes no difference in the results. Static models can be sensitive to the choice of time unit.

Loss estimates provided by the policy model are expected present values, averaging out years in which disruptions do and do not occur according to the specified probabilities. The possibility of multiple disruptions is recognized, and losses are more heavily discounted, the further into the future they occur.

Summary of Parameters and Loss Estimates

To summarize, the key parameters used as inputs by the policy model are as follows: the quantity of the material produced and consumed domestically; the normal world market price; the slopes of the domestic supply and demand curves in the short run and long run; the efficient speed of adjusting capacity in the absence of stock releases; the world market price during disruptions, and decreases therein that specified stock releases would cause; foreign consumers' stocks; the probabilities that a disruption will begin and continue, and decreases therein that specified larger stockpiles would cause; the rate at which future costs and benefits are to be discounted; and the unit cost of stockpiling. In its standard mode, the policy model assumes U.S. tariffs do not reduce the likelihood or severity of market disruptions (in contrast to stockpiles, according to the probability-deterrence and price-deterrence effects); in that case there are optimally no U.S. tariffs and subsidies in the market under study.

The estimate of total national expected economic losses provided by the policy model is broken down into a number of categories:

1. domestic consumers' cost of conservation or substitution;
2. domestic producers' added costs of production;
3. the cost of predisruption capacity preparation in anticipation of disruptions;
4. the cost of rapid capacity adjustment on the supply and demand sides during disruptions;
5. costs of stockholding; and
6. transfers to foreign suppliers.

Furthermore, it is possible to generate information on transfers from consumers in the United States to other U.S. parties, notably, to domestic producers and those holding stocks. These transfers do not affect the standardized calculation of total national economic losses (the disaggregation of which was described above), but they are very likely to make important differences in policy making. Expected economic losses for the nation are calculated assuming the same threat of disruption, in terms of probabilities, severity, and so on, continues indefinitely into the future.

ILLUSTRATIVE APPLICATION OF THE POLICY MODEL TO DETERMINING OPTIMAL CHROMIUM STOCKPILES

As discussed in Chapter 6, chromite is a vital U.S. import used to make stainless steel.[3] The following sample calculations for the chromite market are based on a disruption scenario in which Rhodesian supplies are completely disrupted (as by a civil war); the Soviet Union, Turkey, and several lesser exporters collude effectively to maximize profits by raising export prices; and South Africa is constrained from expanding exports by transportation bottlenecks and limited civil unrest. Other, less severe disruption scenarios are perhaps more probable than this very pessimistic possibility, but such a case is certainly not unthinkable in view of recent events in southern Africa and the appearance since 1975 of some tacit collusion between the Soviet Union and Turkey on chromite prices. Although the sample calculations presented below were in fact based on this scenario, they are primarily presented as a numerical example inasmuch as a complete explanation of all assumptions would be too voluminous.

Parameter values specified for our numerical example are summarized in Table A.1. The quantity of U.S. chromium imports per year is multiplied by its normal price in the form of ore, yielding an annual value of imports of $140 million. (We ignore here certain complications arising from the fact that chromium is imported as ferroalloys and in other more processed forms, though we do include the quantity of chromium imported as ferroalloys in our calculations.) The short-run and long-run price elasticities of demand are respectively -0.02 and -0.07, implying that a 10 percent price increase will cause (approximately) a 0.2 percent decrease in consumption in the short run and a 0.7 percent decrease in the long run. The real interest rate is assumed to be 6 percent per year; it is used both to discount future costs and benefits, and to determine the annual interest cost of holding the equivalent of one year of U.S. consumption in a stockpile for one year: 6 percent of $140 million equals $8.4 million per year. Stocks held by foreign consumers of chromium, available for release during a disruption, are assumed to be equal to 1.5 years of U.S. consumption.

The probability of a disruption beginning during the year following any point in time when there is no disruption is assumed to be 0.04 per year; this "inception probability" is equivalent to a probability of 0.335 (or approximately one chance in three) that a disruption will occur over the course of a decade. When a disruption is in progress, the "continuation probability" of a disruption continuing for at least another year is assumed to be 0.8 per year.

TABLE A.1

Summary of Parameters for Illustrative Chromite Supply-Disruption Scenario

---

Annual value of imports (as ore): $140 million

Demand (point) elasticities: { -0.02 short run; -0.07 long run }

Real interest rate: 6 percent

Annual cost of holding stocks equal to one year of U.S. consumption: (0.06) ($140 million) = $5.8 million

Foreign stocks: equal to 1.5 years of U.S. consumption

Disruption probabilities: { 0.04 inception, per year; 0.8 continuation, per year }

Disruption severity: world price

= 4.5 x normal price in absence of stock release
= 3.0 x normal price, with world releases equal to one year of U.S. consumption

Adjustment in intersection of short-run and long-run demand, in the absence of stock releases:

Years' duration	1	2	4	8
Percent adjustment	10	30	54	93

Source: The numerical example which these parameters imply is discussed in more detail in Charles River Associates, The Report of the U.S. Department of the Interior on the Critical Materials Aluminum, Chromium, Platinum and Palladium: A Review and Revision (Cambridge, Mass.: CRA, 1977), Chapter 6, particularly Table 6.3.

The severity of the disruption is summarized from prior analysis as follows: in the absence of any stock releases, the world price at which U.S. imports can be obtained is 4.5 times the normal price for an undisrupted market. If all consuming nations (including the United States, Western Europe, and Japan) release an amount of stocks per year which is the equivalent of one year of U.S. consumption, then the world price is 3.0 times the normal price. The effect on the world price of smaller and larger stock releases during the disruption is determined by linear interpolation and extrapolation using the two cases just stated. The case in which there are no stock releases, and the market price rises to 4.5 times its normal level, is used to calibrate the speed at which the intersections of short-run and long-run U.S. demand curves will adjust from its predisruption position toward its position of ultimate adjustment (should the disruption last indefinitely long). Two intermediate parameters (not discussed further here) are selected which imply a 10 percent adjustment after one year, a 30 percent adjustment after two years, a 64 percent adjustment after four years, and a 93 percent adjustment after eight years. The costs of accelerated adjustment are deduced from assuming this adjustment pattern is optimal in the absence of stock releases. When stock releases are being analyzed, the optimal speed of adjustment and the corresponding costs are reduced. (See the loss category "Premium payments for rapid adjustments by U.S. consumers during disruptions" in Table A.2 below.) U.S. production of chromium is negligible and is assumed to remain so during disruptions.

Assuming there is one chance in three that a chromite disruption as serious as that described above will start in the course of a decade, the optimal total U.S. stockpile would be equal to over 21 months of normal consumption, according to the optimal policy model described above. Assumptions about the likelihood of disruptions starting and continuing are important determinants of the most efficient stockpile size. For example, if the probability of a disruption's starting in the course of a decade dropped from 0.335 (that is, about one chance in three, as assumed above) to 0.30, then the efficient size of the U.S. stockpile would drop from the equivalent of over 21 months of consumption to 16 months. On the other hand, if this inception probability were 0.40 per decade, then the efficient stockpile would equal almost 31 months of consumption.

If the chromite-disruption threat described above continues for many years (and demand remains constant over time), then U.S. economic losses of $47 million per year can be expected, averaging out years in which disruptions do and do not occur, and

TABLE A.2

Illustrative Chromite Supply-Disruption Scenario: Stockpile Levels and Associated Average Annual U.S. Losses

Loss Category	Socially Optimal Stockpile: 21 Months of Normal Consumption ($ million)	Private Competitive Stockpile: 6 Months of Normal Consumption ($ million)
Average U.S. economic loss per year	47.0	49.3
Breakdown		
Additional payments on remaining imports during disruptions	30.1	42.3
Cost of conservation by U.S. consumers during disruptions	1.7	2.0
Premium payments for rapid adjustments by U.S. consumers during disruptions	0.2	0.3
Cost of adjustments by U.S. consumers after disruptions end	0.4	0.5
Stockholding costs	14.6	4.2
Net losses on stockpile operation	3.5	0

Source: Charles River Associates, The Report of the U.S. Department of the Interior on the Critical Materials Aluminum, Chromium, Platinum and Palladium: A Review and Revision (Cambridge, Mass.: CRA, 1977), Table 6-4, pp. 6-29 through 6-34.

assuming that the economically efficient stockpile (21 months of normal consumption) is held before disruptions begin, and that an optimal release rate is followed. Table A.2 indicates the types of economic losses that make up this average. The United States would continue to import substantial quantities of chromite during disruptions, and additional payments for these remaining imports (beyond what they would cost in normal times) represent the major category of loss: an expected value of $30.1 million per year. U.S. consumers would not cut their consumption a great deal, even if prices for the ore chromite were more than three times normal prices; for this reason, the costs of conservation are relatively small: $1.7 million per year. Adjustments by U.S. consumers would typically be rush jobs after a disruption had begun, involving premium payments roughly estimated at $0.2 million per year. Adjustments made by U.S. consumers during a disruption would, in many cases, be undone after the disruption ended, involving average losses estimated at $0.4 million per year. It is assumed that U.S. production of chromite is not substantial during disruptions, so the costs of expanding (and contracting) U.S. output are not included in the table.

Holding the economically efficient stockpile is quite expensive, averaging $14.6 million per year, according to the last item of the breakdown in the table. Stock sales during disruptions yield revenues averaging $11.1 million per year, for a net loss on stockpiling operations averaging $3.5 million per year. In contrast, the table indicates that private competitive stockpilers would stop stockpiling as soon as it became unprofitable, holding stocks equal to six months of normal consumption instead of 21 months. This distinction between the total private competitive stockpile, which individual firms would find most profitable, and the socially optimal stockpile is based entirely on the price-deterrence effects embodied in the description of the disruption severity in Table A.1. Although stockpiling costs would, of course, be less with small holdings, all other loss categories would be greater, especially the losses that would result from additional payments on remaining imports during disruptions.

For the particular illustrative chromite-disruption scenario under discussion, increasing U.S. stockpiles from six months to 21 months of normal consumption only decreases average yearly U.S. losses from $49.3 million to $47 million. However, if stockpiles decrease the probability that a chromite cartel will start and continue, then the benefit of going from the private competitive stockpile to the social optimum can be much more dramatic.

The reason price-deterrence effects cause a distinction between the socially optimal stockpile and the private profit-maximizing stockpile is illustrated in Figure A.4. Four U.S. firms that normally import chromite from abroad are represented by the small circles on the left. If during a supply disruption the first U.S. firm consumes out of its stockpile (represented by the small triangle) rather than buying on the world market, then the world price is lower and other U.S. consuming firms pay less for their imports. Each U.S. firm must bear the entire cost of holding an additional unit of stocks, but much of the benefit from reduced import costs is external to the individual firm. Thus, the social and private optimums diverge.

FIGURE A.4

U.S. Consumers

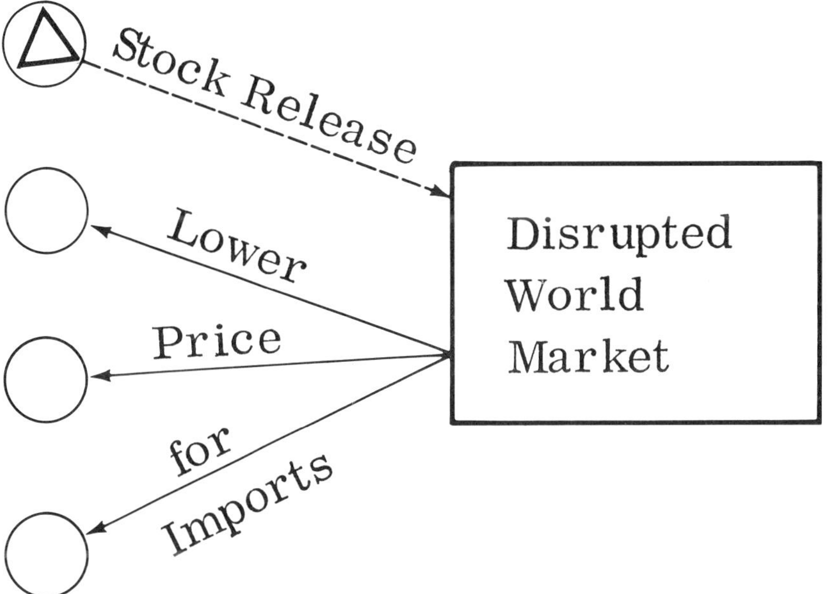

Source: Constructed by the authors.

Probability-deterrence effects operate in a similar fashion to make the socially optimal stockpile larger than the privately optimal stockpile, though assuming larger U.S. stocks would reduce the chance of a disruption is typically most relevant where the disruption would be the result of calculated prior collusion by foreign suppliers. Again, there are external benefits from additional stockholding (above the private optimum) by individual U.S. firms, which accrue to all other consuming firms, that is, a smaller probability of having to endure a disruption at all.

The socially optimal stockpile equaling 21 months' consumption, calculated above, assumed the inception probability was constant at 0.04 per year (equivalent to 0.335 per decade). Suppose, however, as indicated in Table A.3, that increasing U.S. stocks from a figure equaling 21 months' consumption to one of 33 months' consumption would reduce the inception probability to 0.032 per year. How much of that additional stockpiling would then be worthwhile, from the standpoint of minimizing expected national losses ?* In fact, stockpiling equal to 35 months' consumption would then be socially optimal, as indicated in the last column of Table A.3. The inception probability would be lowered all the way to 0.031 per year. The particular probability deterrence effects specified in Table A.3 are presented as a numerical example, not part of any particular disruption scenario. However, the magnitude of probability deterrence which is assumed would be plausible in many cases.

Assuming this relationship between stock size and inception probability, the private profit-maximizing stockpile for competitive firms would equal 14 months' consumption. Table A.4 indicates that expected annual U.S. losses obtained by increasing U.S. stocks from the equivalent of 14 months' consumption to that of 35 months' consumption are $43.9 million versus $52.6 million, a much more substantial difference than was the case with price-deterrence effects alone. It is remarkable to note that stockholding costs become the largest loss category under these assumptions. Imports during disruptions are relatively small because of the possibility of massive stock releases.

---

*A log-linear relationship between U.S. stock size and the inception probability is automatically fitted by the policy model to the two data points given in the first and second columns of Table A.3.

TABLE A.3

Relationships between the U.S. Ratio of Stocks to Consumption and the Inception Probability (deterrence to inception probability)

	Previous Social Optimum	Effect of Hypothetical Stock Increase	New Social Optimum
U.S. stocks/consumption	21 months	33 months	35 months
Inception probability	0.04	0.032	0.031

Source: Charles River Associates, The Report of the U.S. Department of the Interior on the Critical Materials Aluminum, Chromium, Platinum and Palladium: A Review and Revision (Cambridge, Mass.: CRA, 1977), Tables 6-4 (a-f), pp. 6-29-6-34.

Results are even more striking when larger U.S. stocks are assumed to reduce the continuation probability as well as the inception probability. Suppose, as indicated in Table A.5, that increasing the U.S. stockpile from stocks equal to 35 months' consumption to those equal to 47 months' consumption would lower the continuation probability from 0.8 per year to 0.75 per year (in addition to further lowering the inception probability, according to the relationship specified in Table A.3). How much of the additional stockpiling would then be worthwhile? In fact, 44 months' consumption would then be the socially optimal stockpile, pushing the continuation probability down to 0.76 per year (and the inception probability down to 0.026 per year).

Average U.S. economic losses per year, under the new assumption about continuation-probability deterrence, are given in Table A.6. The benefits of moving from the private competitive stockpile to the socially optimal stockpile are now quite large, but stockholding costs are massive. Very large stocks are held to reduce disruption probabilities, that is, reduce the chance that the stockpiles will in fact be used. Thus, net losses on stockpile operations are very large.

TABLE A.4

Illustrative Chromite Supply-Disruption Scenario: Stockpile Levels and Associated Average Annual U.S. Losses with Price and Inception Deterrence

Loss Category	Socially Optimal Stockpile: 35 Months of Normal Consumption ($ million)	Private Competitive Stockpile: 14 Months of Normal Consumption ($ million)
Average U.S. economic loss per year	43.9	52.6
Breakdown		
Additional payments on remaining imports during disruptions	17.7	40.5
Cost of conservation by U.S. consumers during disruptions	1.2	2.1
Premium payments for rapid adjustments by U.S. consumers during disruptions	0.2	0.3
Cost of adjustments by U.S. consumers after disruptions end	0.3	0.5
Stockholding costs	24.5	9.2
Net losses on stockpile operation	12.8	0

Source: Charles River Associates, The Report of the U.S. Department of the Interior on the Critical Materials Aluminum, Chromium, Platinum and Palladium: A Review and Revision (Cambridge, Mass.: CRA, 1977), pp. 6-29 through 6-34.

TABLE A.5

Relationships between the U.S. Ratio of Stocks to Consumption and the Probabilities of Inception and Continuation for a Disruption (deterrence to inception and continuation probability)

	Previous Social Optimum	Effect of Hypothetical Stock Increase	New Social Optimum
U.S. stocks/consumption	35 months	47 months	44 months
Inception probability	0.031		0.026
Continuation probability	0.8	0.75	0.76

Source: Charles River Associates, The Report of the U.S. Department of the Interior on the Critical Materials Aluminum, Chromium, Platinum and Palladium: A Review and Revision (Cambridge, Mass.: CRA, 1977), Tables 6-4 (a-f), pp. 6-29 - 6-34.

STOCKPILE RANGES FOR CHROMITE

Analysis of disruptions in the chromite market, such as the example presented in detail above, suggest that a contingency stockpile of chromium can be easily justified under pessimistic assumptions regarding future contingencies. Price-deterrence effects provide part of the rationale for this recommended stockpile. Events in southern Africa that are most likely to initiate a major disruption in the chromite market would probably not be deterred by large U.S. stockpiles. However, subsequent collusion among the remaining producers, notably, the Soviet Union and Turkey, might well be discouraged by larger U.S. stocks. Thus, probability-deterrence effects also provide part of the rationale for the recommended stockpile.

ECONOMICALLY EFFICIENT STOCKPILE RANGES FOR OTHER MATERIALS

The policy models discussed above have been applied to most of the other materials treated in this book. The remainder of this appendix discusses these analyses for the optimal stockpile ranges summarized previously.

TABLE A.6

Illustrative Chromite Supply-Disruption Scenario: Stockpile Levels and Associated Average Annual U.S. Losses with Price, Inception, and Continuation Deterrence

Loss Category	Socially Optimal Stockpile: 44 Months of Normal Consumption ($ million)	Private Competitive Stockpile: 18 Months of Normal Consumption ($ million)
Average U.S. economic loss per year	41.7	64.5
Breakdown		
Additional payments on remaining imports during disruptions	9.1	48.7
Cost of conservation by U.S. consumers during disruptions	0.7	3.2
Premium payments for rapid adjustments by U.S. consumers during disruptions	0.1	0.4
Cost of adjustments by U.S. consumers after disruptions end	0.2	0.7
Stockholding costs	31.6	11.6
Net losses on stockpile operation	21.9	0

Source: Charles River Associates, The Report of the U.S. Department of the Interior on the Critical Materials Aluminum, Chromium, Platinum and Palladium: A Review and Revision (Cambridge, Mass.: CRA, 1977), pp. 6-29 through 6-34.

Petroleum

Our analysis of possible disruptions in the petroleum market has concentrated on scenarios involving relatively short-lived disruptions, which are, however, relatively likely over the course of a decade. These disruptions are assumed to be similar in expected duration to the OPEC embargo of 1973, but differ in that termination of the embargo results in a drop in petroleum prices to predisruption levels. The resulting socially optimal stockpiles equal the order of magnitude of those recommended in Table 4.4, that is, roughly two months of U.S. consumption.

The assumptions leading to these results include the following:

1. The United States normally imports petroleum equal to 40 percent of its consumption;
2. A 37 percent effective embargo would increase prices two and a half times in the absence of stock releases;
3. If stock releases completely cover the deficit in imports, domestic prices remain at the predisruption level;
4. The probability of a disruption's starting over the course of a year is 0.06 (0.46 per decade);
5. The probability of a disruption's continuing as long as a year, once it has begun, is 0.3 (0.09 for two years; 0.55 for six months);
6. If U.S. stocks were equal to slightly more than one year of U.S. imports rather than four months of imports, the probability of a disruption's starting in the course of a year would be lowered from 0.06 to 0.04;
7. For the stock increases in 6 above the probability of a disruption's continuing for as long as a year would be lowered from 0.3 to 0.2.

The calculations are based solely on direct costs inflicted on those U.S. industries and consumers who consume petroleum and petroleum products, and do not include indirect costs stemming from macroeconomic or political effects.

The results from our policy model are again particularly interesting as an illustrative case in which private U.S. industry would have sufficient incentive to hold only a small fraction of the socially optimal stockpile. As discussed above in the case of chromite, the external benefits from price-deterrence effects are inadequately valued by the individual stockpiler. An efficiently run government stockpile could, in the face of the threat described in detail above, reduce average U.S. economic losses per year from over $2 billion to roughly two-thirds of that amount. We do not consider here the difficult issue of coordinating government stockpiling with private inventory holding.

We have not performed a full-scale analysis of petroleum stockpiling for foreign supply disruptions using the generalized policy model, including either more severe but less frequent disruptions or indirect costs. However, it appears probable that a contingency stockpile equaling perhaps two to three months of U.S. consumption would be justified under moderate assumptions.

Manganese and Chromium and the Specification of Disruption Probabilities

Not surprisingly, for manganese or any other threatened material, the specified probabilities of a disruption's beginning and continuing have a very strong effect on the efficient stockpile size for the United States. Choosing the appropriate probabilities, in conjunction with specifying the severity of the disruption scenario or scenarios to be considered, is a difficult task requiring considerable judgment. In most cases such probabilities cannot be estimated with much confidence from a simple examination of the history of the particular market under consideration.

One approach to this problem is to examine the trend level of, say, manganese stocks, presumably as a proxy for contingency stocks held by U.S. consumers, and to assume that these stocks are efficient for the private sector, given their perceptions of the risks they face. The probabilities specified for the policy model can then be adjusted to yield an optimal private competitive stockpile approximately equal to that which is actually observed (assuming that the relevant industries behave competitively). Because private competitive stockpiles are treated as given data in this approach, attention shifts to the relative size of efficient stocks from a public point of view.

For both manganese and chromite, an examination of historical industry stock levels, as estimated by the U.S. Bureau of Mines, suggests that contingency stocks equaling roughly six months' consumption are held because these materials are imported over long supply lines that are subject to disruptions. Total industry stocks have tended to equal more than six months' consumption at times when they could be adjusted to desired levels, but stocks equal to several months of consumption must be allocated to normal working inventories, which would be necessary even if the material were obtained from very secure domestic sources. The efficient stocks calculated by the policy model correspond most closely to the desired excess of stocks over normal working inventories.

The approach outlined above was in fact followed to a significant extent when the calculations (presented earlier) for chromite were done. Thus, attention centers on how much larger the socially optimal stockpile is relative to the private optimum. For both manganese and chromite, typical results for socially optimal stocks show that sizes range from two to five times the size of private competitive stocks. (The examples for the chromite market presented earlier lie in the middle of this range.) As noted earlier, the policy model presents a very strong rationale for government stockpiling, though, in the final analysis, the benefits of a government stockpile must be weighed against the difficulties, political and otherwise, of managing it in a manner approaching the perfect efficiency assumed by the model.

In the mid-1970s, the threat of a major disruption of U.S. manganese imports appeared to be somewhat less severe than threats to chromite imports. Thus, based on limited analyses completed to date, a nonmilitary contingency stockpile equal to 12 to 18 months of U.S. manganese consumption appeared justifiable.

The specified size of manganese stocks held by foreign consumers can have a stronger impact on efficient U.S. policy than might be immediately appreciated. Since an international market model underlies specification of the price-deterrence effect in the policy model, smaller stocks held by foreign consumers imply higher world prices during a disruption and, hence, larger efficient U.S. stockpiles. In the usual cases, socially optimal U.S. stocks are less affected by changes in the stocks held by foreign consumers than are optimal private stocks.

The optimal policy model does not address the issue of how larger U.S. government stockpiles might discourage private stockpiling or other preparations for disruptions. The private competitive optimum is calculated assuming no government stock releases. The calculated socially optimal stockpile corresponds to the sum of public and private stocks, with no indication of the relative sizes or interactions of these two components. The logic of the policy model would generally imply that, where large government stocks are held, the private sector would only hold sufficient stocks to last from the beginning of a disruption until government stocks would be released; there is a strong incentive for private firms to pursue such a policy if they wish to minimize their expected losses from disruptions and from the costs of preparing for disruptions (particularly, the cost of stockpiles).

In the past there was considerable uncertainty about the availability of U.S. government stocks of manganese and other materials for any event short of a military emergency. This uncertainty probably induced U.S. consumers to hold larger stocks than would have been the case if they knew stock releases could readily be triggered by nonmilitary market disruptions. During the 1960s, decreases in military stockpile objectives, and development of a general pattern of stock releases by the U.S. government, signaled the greater availability of government stocks to compensate for nonmilitary supply disruptions such as the embargo of Rhodesian chromite.* However, such a means of signaling availability of stocks is inherently self-limiting when the stockpile reaches specified military objectives and there is no substantial excess. In 1976 the stockpile objectives were again raised; on the other hand, there was serious consideration given in the government to policies to make government stocks available as a tool to counteract embargoes and cartel formation by foreign suppliers. On net balance, it is probably the case that during the 1970s private firms have lowered their expectations of government stockpile assistance in the event of a peacetime emergency.

Zairian Cobalt and Exogenous Political Events

The analysis of supply disruptions in this volume has focused on deliberate policy decisions of exporting countries, particularly decisions to join with other exporters in a cartel or in an embargo of the United States. In 1977 and 1978, exports of cobalt and copper from Zaire were seriously disrupted by invasions of Shaba Province by forces based in neighboring Angola. Although such a disruption is not a deliberate embargo, its effects on the cobalt market are similar, and much of our analysis of deliberate supply disruptions in Chapter 8 is directly relevant.

In the spring of 1977, shortly after the invasions of Shaba Province began, CRA analyzed the possible implications for the cobalt market.[4] Part of this study involved applying our policy model to determine optimal cobalt stockpiles, assuming the threat to Zaire's exports continued, but stocks could be accumulated sufficiently slowly so that only normal prices would have to be paid for the accumulated cobalt. A possible alternative source of cobalt would be the U.S. strategic stockpile, but its availability for

---

*It may be that during the 1960s private stockholders would have held contingency stocks of manganese and chromite equal to significantly more than the six months' consumption discussed above if there had been no government stockpile. If so, that earlier analysis should in principle be adjusted.

nonmilitary emergencies had, shortly before, become more doubtful when proposed objectives for military contingencies were raised considerably above existing levels of stocks.

Application of the policy model to the cobalt market presented an interesting contrast to other applications discussed above. When cartel formation is the item of interest, the market goes from a competitive structure in normal times to a monopolistic structure during the disruption. However, we have just the opposite situation for cobalt. During normal times, Zaire exercises considerable monopoly power in its pricing and output decisions. If supplies from Zaire are completely disrupted, no other producer has sufficient market power to take Zaire's place; a coalition of the remaining producers is very unlikely.

Economically efficient stockpiles calculated by the policy model, for a complete disruption of Zaire's supplies from the world market, are not greatly different from those reported in Chapter 8, based on examination of deterrence effects and disruption probabilities that would have been reasonable before the invasion of Shaba Province. Socially optimal stockpiles are somewhat larger, but not nearly as much as was the case for chromite or manganese. The reason is that price-deterrence effects and probability-deterrence effects do not operate strongly for the type of disruption we have been assuming for the cobalt market.

Of course, given the major disruption in the cobalt market in 1978, with the dealer's price approaching ten times the predisruption level, it would, in retrospect, be easy to justify a huge cobalt stockpile. However, calculation of such an efficient stockpile could only be of academic interest, since its acquisition at predisruption prices (as assumed by the policy model) is impossible. Our above discussion of optimal contingency stocks in the cobalt market has assumed a significant sustained threat, with an inception probability on the order of 0.08 per year; over a five-year period, this is equivalent to an inception probability of 0.34.

Aluminum/Bauxite

Application of our policy model to the aluminum/bauxite market is less straightforward than was the case for the materials considered above. The primary reason is that there are massive aluminum resources in the United States and elsewhere that would very probably be exploited if the price of bauxite delivered from Jamaica to the United States were to double or triple over the level of the mid-1970s. At least one U.S. aluminum company has claimed that exploitation of nonbauxitic U.S. resources could be made profitable if bauxite prices rose only moderately. For this reason, our basic results were obtained using a conceptually identical but factually specific model.

Under these circumstances, a single price-deterrence relationship, relating the price of U.S. imports to the amount of U.S. stock releases, will be inadequate if bauxite prices remain sufficiently high for sufficiently long during a disruption. Also, considering the fact that bauxites from different sources are imperfect substitutes and that differences in transportation costs to the United States are significant, it is clear that the United States is in a bargaining situation vis-à-vis the nearby Caribbean bauxite producers, which have traditionally supplied this country. For example, by imposing a tariff on bauxite imports sufficient to make domestic resources a clearly competitive substitute, the United States could impose heavy losses on foreign bauxite suppliers; the higher the price of imported bauxite, the lower the cost of such a policy relative to its potential benefits. Our policy model in its current form does not allow analysis of such bargaining ploys. In Chapter 5 these possibilities are considered without formal models of the bargaining process.

Our policy model has been applied both to scenarios involving severe short-run price gouging by foreign bauxite producers, and to scenarios involving sustained doublings of prices that are assumed not to induce massive additional exploitation of U.S. aluminum resources. Economically efficient stockpiles similar to those reported in Chapter 5, equaling consumption ranging from six months to a year, are obtained for reasonable parameter values. However, for reasons discussed above, this type of analysis is too narrowly focused to be a fully reliable policy guide. Chapter 5 considers a wider range of disruption scenarios, though the effects of deterrence benefits on optimal stockpile size are not considered with the sophistication that is possible with our generalized optimal policy model.

One conclusion which emerges quite clearly from application of our policy model is that economically efficient stockpiles are considerably larger than private competitive stockpiles for any disruption scenario that does not involve massive displacement of imports by domestic mining of nonbauxitic ores. Probability-deterrence effects for stockpiles are very plausible, particularly for short-run price-gouging scenarios, and price-deterrence effects operate through imports during the disruption. In a bargaining model, national coordination of stockpiling policy would presumably also be necessary if it were to be used as a bargaining tool along with such ploys as threatening a tariff on imports of bauxite.

Platinum and Palladium

The platinum and palladium markets differ from those considered earlier in that secondary recovery is very important, particularly for catalysts and other end uses where the material is clearly a stock in use that is refined year after year, with only small dissipation of the original amounts. As discussed in Chapter 9, the unpredictability of speculative stockholding reduces the usefulness of optimal stockpile analyses based on rational assumptions regarding speculation. These analyses lead to calculation of smaller economically efficient stockpiles than might be needed to offset the impacts of speculative hoarding.

Our general policy model, as applied to other materials in this appendix, has not yet been generalized to incorporate fully the possible benefits of expanded secondary recovery during disruptions, so we modified and simplified it in some ways to permit its application to markets with large stocks in use. One interesting issue is whether consumption of a threatened imported material should be encouraged or discouraged during normal times. The usual argument is that consumption should be discouraged so that users have less costly adjustments to make after a disruption begins. For materials such as chromium or aluminum, secondary recovery of obsolete scrap is significant, though much less than 10 percent of market demand in typical years. It is conceivable that consumption of such materials should be encouraged in normal times in order to facilitate secondary recovery during disruptions, though certainly such an effect operates very much less powerfully than in the platinum market, where recovery rates over 90 percent are common.

Copper

The version of the policy model described in this appendix has not yet been applied to the copper market. However, since the United States would move strongly toward self-sufficiency during a serious disruption of imports, price-deterrence effects would not be a dominant consideration. The importance of probability-deterrence effects is problematical, given the difficulty in specifying realistic scenarios of serious disruptions. In the absence of these deterrence effects, the analysis of copper stockpiles reported in Chapter 10 requires no extensions here.

NOTES

1. For a full discussion, see Charles River Associates, _The Report of the U.S. Department of Interior on the Critical Materials: Aluminum, Chromium, Platinum and Palladium: A Review and Revision_ (Cambridge, Mass.: CRA, July 1977).

2. The revised policy model, and its application to the markets for chromium and aluminum, are presented in ibid. The studies of platinum and palladium utilized a simplified and specialized policy model that is discussed briefly near the end of this appendix.

3. Results presented below are discussed more fully in ibid.

4. See Charles River Associates, _Implications of the War in Zaire for the Cobalt Market_, prepared for the Office of Minerals Policy and Research Analysis of the U.S. Department of the Interior. The revised edition was dated June 1977; it took into account information provided by U.S. cobalt suppliers in response to the initial edition issued several months earlier.

INDEX

ABOUT THE AUTHORS

MICHAEL KLASS is assistant director for economic evidence of the U.S. Federal Trade Commission. Before joining FTC, he taught at the University of Michigan, was a senior research associate at Charles River Associates, and served as staff economist for regulatory reform for the U.S. Senate Governmental Affairs Committee. At CRA he directed a study of material supply problems and developed a conceptual framework for analyzing cartel behavior and the benefits and costs of alternative government policies, such as tariffs and quotas. He was also involved in studies of economic interdependence between the United States and suppliers of foreign commodities, and competition in the electric power industry. His publications include Regulation and Entry (with W. Shepherd) (East Lansing, Mich.: Michigan State University, 1974); and A Framework for Regulation (with L. W. Weiss), U.S. Senate Governmental Committee (Washington, D.C.: Government Printing Office, 1978).

Dr. Klass received his Ph.D. in economics from the University of Wisconsin.

JAMES BURROWS is vice president and director of the Metals and Minerals Program at Charles River Associates. An expert on natural resource-based industries, he has directed studies of the metals and minerals industries for more than ten years and has published extensively in the field. He has supervised studies on all of the major commodities in the national stockpile, the likelihood and impacts of supply interruptions in strategically important minerals markets, and alternative policy options open to the U.S. government for dealing with cartel formation. In addition, he has supervised studies on international commodity agreements in metals and minerals markets and on the effects of possible future commodity shortages. He is the author of the following books: (with Thomas A. Domencich) An Analysis of the U.S. Oil Import Quota (Lexington, Mass.: Heath-Lexington Books, 1970); Cobalt: An Industry Analysis (Lexington, Mass.: Heath-Lexington Books, 1971); Tungsten: An Industry Analysis (Lexington, Mass.: Heath-Lexington Books, 1971); (with Charles Metcalf and John B. Kaler) Industrial Location in the United States (Lexington, Mass.: Heath-Lexington Books, 1971); (with Douglas Woods) Aluminum and the IBA (New York: Praeger, 1978).

Dr. Burrows received his Ph.D. in economics from the Massachusetts Institute of Technology.

STEVEN BEGGS is a senior research associate at Charles River Associates. In addition to his work on analyzing the criticality of commodities in the national stockpile, and evaluating government policy alternatives in dealing with potential producer-country cartelization of strategic supplies, he has conducted research on forecasting the demand for electric automobiles and on the world rubber, copper, cobalt, chromium, and manganese markets. He has also analyzed long-run supplies of less common materials which may in the future be consumed in much larger quantities for new technologies such as advanced storage batteries.

Dr. Beggs received his Ph.D. in economics from the Massachusetts Institute of Technology.

ABOUT CHARLES RIVER ASSOCIATES

Charles River Associates specializes in applied microeconomic analysis for industry, government, and nonprofit organizations. Since its founding in 1965, CRA has conducted research in:

commodity-market forecasting
antitrust policy
combined economic engineering feasibility for new ventures
communications
consumer behavior
economic development
environmental economics
fuel industries, electric power, and energy economics
industry regulation
international trade
minerals, metals, and durable-goods industries
regional economics
transportation planning
urban and intercity transportation economics

F